普通高等职业教育计算机系列规划教材

Java EE 实例开发项目教程

（Struts2+Spring+Hibernate）

李明欣　康　凤　主　编

冯海波　林　琳　朱卫平　陈　蕾　副主编

王　津　主　审

電子工業出版社

Publishing House of Electronics Industry

北京 • BEIJING

内容简介

本书是作者多年来实践经验的总结，汇集了作者在教学和项目中遇到各种问题及解决方案。

本书采用迭代的方式讲解，以实际项目中的增删改查（CURD）为基础，采用不同的框架技术迭代实现，在这个过程中首先引入 Struts2，接着讲解 Struts2 相关技术，最后使用 Struts2+JDBC 实现增删改查，接着引入 Spring 框架，接着讲解 Spring 框架相关技术，最后使用 Struts2+Spring+JDBC 实现增删改查，最后引入 Hibernate 框架，首先讲解 Hibernate 相关技术，最后使用 Struts2+Sprng+Hibernate 实现增删改查。在学习完 Struts2+Spring+Hibernate 三大框架之后，最终达到整合三大框架开发实际项目，并在最后一章提供了综合案例-博客系统。本书是一本注重实际操作的实例教程，在讲解的过程中，只需掌握基本的理论，然后通过实战案例就能轻松掌握。

本书内容全面，结构清晰，注重实战，非常适合 Java Web 开发人员学习使用，同时也可以作为软件公司的参考书。

图书在版编目（CIP）数据

Java EE 实例开发项目教程：Struts2+Spring+Hibernate/李明欣，康凤主编. —北京：电子工业出版社，2016.3
ISBN 978-7-121-27892-1

Ⅰ. ①J… Ⅱ. ①李… ②康… Ⅲ. ①JAVA 语言－程序设计－高等学校－教材 Ⅳ. ①TP312

中国版本图书馆 CIP 数据核字(2015)第 307524 号

策划编辑：徐建军（xujj@phei.com.cn）
责任编辑：郝黎明
印　　刷：三河市鑫金马印装有限公司
装　　订：三河市鑫金马印装有限公司
出版发行：电子工业出版社
　　　　　北京市海淀区万寿路 173 信箱　邮编 100036
开　　本：787×1 092　1/16　印张：17.25　字数：441.6 千字
版　　次：2016 年 3 月第 1 版
印　　次：2016 年 3 月第 1 次印刷
印　　数：3 000 册　　定价：38.00 元

凡所购买电子工业出版社图书有缺损问题，请向购买书店调换。若书店售缺，请与本社发行部联系，联系及邮购电话：（010）88254888。

质量投诉请发邮件至 zlts@phei.com.cn，盗版侵权举报请发邮件至 dbqq@phei.com.cn。

服务热线：（010）88258888。

前　言

本书以“突出技术，强化能力，综合应用”为指导思想，通过项目驱动，多个实际应用案例为引导，介绍 Struts2+Spring+Hibernate 三大框架技术。书中的技术介绍、实际项目载体以及经典模块代码都来源于编者多年的教学与开发工作积累，具有实践性和操作性，从而帮助读者尽快掌握这些技术框架的使用方法。本书的特色如下：

1．内容编排由浅入深，知识点新颖，紧跟时代潮流

本书章节按照由浅入深、循序渐进的顺序编排。内容选取上，精选 Struts2、Spring、Hibernate 框架中的流行技术，理论知识坚持够用原则，项目选择注重实用性和代表性。书中介绍的具体案例实现过程，也是培养读者实际参与项目开发的能力，适应社会，技术提高的过程。

2．实例丰富，具有实操性

本书每章的知识点都配备了大量的应用实例，这些实例充分展现了相关知识点的实现细节。读者可以在学习相关知识点后，结合实践来更形象深入地了解并运用这些知识点。案例的实现，也是知识融会贯通的过程。

3．提供了完整的项目实例，培养读者综合应用能力

本书以当前 Java EE 技术的主流开发技能需求出发，前台页面采用 BootStrap 框架，结合使用 Struts2+Spring+Hibernate 技术，实现了完整的博客系统。通过对项目深入研究，读者可以比较全面地掌握基于 Java EE 应用程序的开发步骤和开发方法，并可将实例中所采用的技术迁移应用到自己的项目中。

4．配有源代码与教学视频，辅助读者自学

为了方便读者掌握本书提供的实例，所有的源代码共享到百度云。另外还提供了高清教学视频，将书中的一些相关操作直观地展示给读者，以帮助读者达到更好的学习效果。

本书以项目驱动方式讲解如何使用 Struts2、Spring、Hibernate 三大框架开发 Java EE 应用程序。第 1 章介绍 Struts2 框架以及开发入门；第 2 章讲解 Struts2 的核心组件，包括 Struts2 的动作、拦截器、注解和 OGNL（对象导航语言）；第 3 章讲解 Struts2 的视图标签，包括数据标签、控制标签、UI 标签和其他标签；第 4 章讲解 Struts2 的国际化和数据校验；第 5 章讲解 Struts2 的应用，包括使用 Struts2 实现文件上传，使用 Struts2+JDBC 实现增删改查；第 6 章讲解 Spring 框架与开发入门；第 7 章讲解 Spring 的 IoC 容器；第 8 章讲解 Spring 的 Bean 管理与配置；第 9 章讲解 Spring（AOP）面向切面的编程技术，第 10 章讲解 Spring JDBC 的使用；第 11 章讲解 Spring 的事务管理；第 12 章讲解 Spring 的应用，包括整合 Struts2+Spring+JDBC 实现增删改查和分页；第 13 章主要讲解 Hibernate 的框架与开发入门；第 14 章讲解 Hibernate 核心 API 的使用；第 15 章讲解 Hibernate 的映射；第 16 章讲解 Hibernate 的查询，包括 HQL 查询、QBC 查询和 SQL 查询；第 17 章讲解整合 Struts2+Spring+Hibernate 实现增删改查和分页，第 18 章主要讲解综合项目，通过项目将前面的知识点全部整合。

本书由成都航空职业技术学院的老师组织编写，由李明欣、康凤担任主编，由成都工业学院的冯海波和成都航空职业技术学院的林琳、朱卫平、陈蕾担任副主编，由成都航空职业技术学院的王津担任主审，另外，参加编写的还有蒋小惠和西华大学的黄曾喜等人。

本书编写过程中参考了众多的技术开发门户网站，包括 www.csdn.net、www.iteye.com、www.github.com 等，并汲取了多方人士的宝贵经验，在此向这些文献的作者和给予帮助的同仁

们表示感谢。

为了方便教师教学，本书配有源码以及相关的学习视频，请有此需要的老师登录华信教育资源网（www.hxedu.com.cn）注册后免费进行下载，还可以到作者的百度云（http://pan.baidu.com/share/home?uk=3473194016）下载。如有问题可在网站留言板留言或与电子工业出版社联系（E-mail：hxedu@phei.com.cn）。

教材建设是一项系统工程，需要在实践中不断加以完善及改进，书中难免存在疏漏和不足，恳请同行专家和读者给予批评和指正。

编　者

2016 年 1 月

目录

Contents

第1章 Struts2 框架与入门

1.1 Struts2 框架

Apache Struts2 是在 Struts1 的基础上注入了 WebWork 的先进设计理念，汲取 Struts1 的诸多优点后而成的一个全新的、非常优秀的 Web 应用程序框架。实践证明 Struts2 能够很好地完成常见领域任务。

1.1.1 Struts2 MVC

MVC（Model-View-Controller）是为了让应用程序开发和维护的效率更高而出现的一种架构型模式。该模式将应用程序划分成模型、视图和控制器三部分，三部分功能明确且相互分离，它改善了应用程序的架构，提高了应用的灵活性和重用性。Struts1 的出现就是把 MVC 模式从桌面应用程序引入 Web 应用程序，这个合成的模式有时也被称为 Model2 模式。这是高效率开发 Web 应用程序的一大进步。而 Struts2 是实现 MVC 设计模式的第二代轻量级 Web 应用程序框架。它致力于将 Web 应用领域的普遍问题或常见情况进行抽象，帮助开发人员快速完成 Web 应用开发。

尽管 Struts1 以其可靠性和稳定性得到业界的认可，但随着计算机软件技术的进步，Struts1 的局限性也逐渐显露出来。Struts2 就是在充分吸取了 Struts1 实践中的经验和教训，提供更加整洁的 MVC 实现而诞生的。同时，它还引入了几个新的架构特性，从而使这个框架更加清晰、更加灵活。这些新特性包括：

- 拦截器：用来从动作逻辑中分层出横切关注点。
- 注解（Annotation）的配置方式：为了减少或者消除 XML 配置。
- OGNL（Object—Graph Navigation Language，对象图导航语言）：一个贯穿整个框架的强大的表达式语言，对集合和索引属性提供强大的支持。

● 标签 API：支持可变更和可重用 UI 组件标签。

框架的出现就是为了帮助我们快速开发，提高效率。它一方面尽量将 Web 应用程序开发过程中的常见问题自动化，另一方面尽量提供优秀的架构解决方案来优化 Web 应用程序中常见的工作流。

Struts2 的高层设计遵循公认的 MVC 设计模式。MVC 模式提供分层结构非常适用于 Web 应用程序，它可以将表现逻辑和业务逻辑分离，构建可复用的软件系统框架，从而简化软件开发。MVC 设计模式中的三部分模型、视图和控制器，在 Struts2 中，它们分别通过动作（Action）、结果（Result）和过滤分配器（FilterDispatcher）来实现。图 1-1 展示了 Struts2 的 MVC 如何处理 Web 应用程序的工作流。通过图示，我们再进一步通过几个名词的对应来学习 Struts2 是如何与 MVC 对应的。

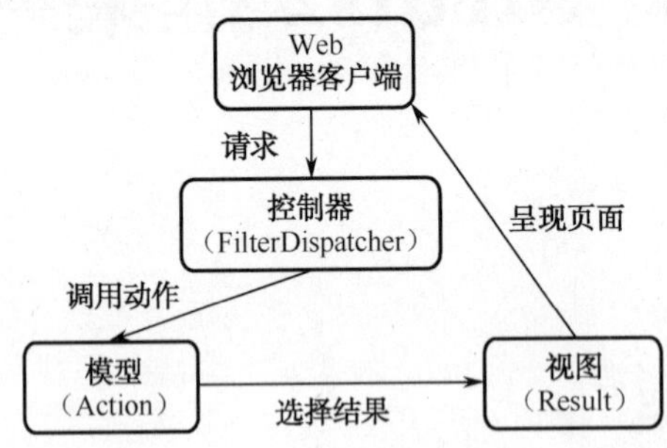

图 1-1　Struts2 MVC 通过 3 个核心框架组件实现：动作、结果和过滤分配器

1. 控制器——FilterDispatcher

Struts2 中使用的 MVC 变体经常被称为前端控制器（Front Controller）MVC。如图 1-1 所示控制器在最前端，是请求处理过程中第一个被触发的组件，其工作是将请求映射到动作。在 Web 应用程序中，传入的 HTTP 请求可以被视为用户向 Web 应用程序发送的命令。Web 应用程序的一个基本任务是将这些请求路由到 Web 应用程序中需要执行的一系列动作。控制器的作用就像是交通警察或者空中交通管制员。从某些方面来看，这个工作具有管控性质，但又不是核心业务逻辑的一部分。

控制器的角色是 Struts2 的 FilterDispatcher 来扮演的。这个重要的对象是一个 Servlet 过滤器，它检查每一个传入请求，决定由哪个 Struts2 动作来处理这个请求。框架帮助你完成所有控制器的任务，你只需要告诉框架哪个请求 URL 需要映射到哪个动作即可。可以通过基于 XML 的配置文件或者 Java 注解来完成这个任务。

2. 模型——动作

模型由 Struts2 动作组件实现。模型是应用程序的内部状态，这个状态由数据模型和业务逻辑共同组成。

Struts2 的动作组件有两个作用。首先，一个动作将业务逻辑调用封装到一个单独的工作单元中。其次，动作是一个数据传输的场所。如图 1-1 所示，控制器在收到请求之后，必须通过映射来决定哪个动作处理这个请求。一旦找到了适当的动作，控制器会调用这个动作并将请求处理的控制权交给它。由框架负责管理的调用过程既准备必要的数据又执行动作的业务逻辑。动作完成它的工作之后，就该向提取请求的用户返回视图了。为此，动作会将结果转发到 Struts2 视图组件。

3. 视图——结果

视图是 MVC 模式的呈现组件。回顾图 1-1 可以发现，动作将获得的数据结果以 Web 浏览器页面形式返回。这个页面是用户界面，向用户呈现应用程序的状态。它们通常是 JSP 页面、Velocity 模板或者用其他表示层技术呈现的页面。虽然视图有多种选择，但是视图的作用很清楚——将应用程序的状态转换为一种用户可以与之交互的可视化的表示。

1.1.2 Struts2 的工作原理

Struts2 提供了更整洁的 MVC 实现。这依赖于其他几个参与每一个请求处理的关键架构组件的帮助。这些架构组件主要包括 ActionContext、Interceptor、OGNL 和 ValueStack。

ActionCentext：是 Action 执行时的上下文。上下文可以看作一个容器（其实我们这里的容器就是一个 Map 而已），它存放的是 Action 在执行时需要用到的对象，比如在使用 WebWork 时，我们的上下文存放有请求的参数（Parameter）、会话（Session）、Servlet 上下文（ServletContext）、本地化（Locale）信息等，在每次执行 Action 之前都会创建新的 ActionContext，ActionContext 是线程安全的，也就是说在同一个线程里 ActionContext 里的属性是唯一的，这样一个 Action 就可以在多线程中使用。

Interceptor：即拦截器，实质就是 AOP（Aspect Oriented Programming，面向切面编程）的一段代码。拦截器允许在 Action 处理之前，或者 Action 处理结束之后，插入执行开发者自定义的代码。

拦截器体系是 Struts2 框架的核心，可以将其理解为一个空容器，大量内建的拦截器完成了该框架的大部分操作。

OGNL：OGNL 是 Object-Graph Navigation Language（对象图导航语言）的缩写。它是一种功能强大的表达式语言，它被集成在 Struts2 框架中用来实现数据转移和类型转换。

ValueStack：ValueStack 实际上就是一个对 OGNL 的封装容器，OGNL 主要的功能就是赋值与取值，Struts2 正是通过 ValueStack 来进行赋值与取值的。

一个请求在 Struts2 框架中的处理过程大概分为以下几个步骤，如图 1-2 所示。

（1）客户端提交一个（HttpServletRequest）请求。

（2）请求被提交到一系列的过滤器（Filter），如 ActionContextCleanUp、其他过滤器（SiteMesh 等）、FilterDispatcher。需要注意的是过滤器的选择是有顺序的，先 ActionContextCleanUp，再其他过滤器（Other Filters、SiteMesh 等），最后到 FilterDispatcher。

（3）接着 FilterDispatcher 被调用，它是控制器的核心，即 MVC 中控制层的核心。它会向 ActionMapper 询问是否需要调用某个 Action 来处理这个（HttpServlet Request）请求，如果 ActionMapper 决定需要调用某个 Action，FilterDispatcher 则把请求的处理交给 ActionProxy。

（4）ActionProxy 通过 Configuration Manager（struts.xml）询问框架的配置文件，找到需要调用的 Action 类。例如，用户注册示例将找到 RegisterAction 类。

（5）ActionProxy 创建一个 ActionInvocation 实例，同时 ActionInvocation 通过代理模式调用 Action。但在调用之前，ActionInvocation 会根据配置加载 Action 相关的所有 Interceptor（拦截器）。

（6）一旦 Action 执行完毕，ActionInvocation 负责根据 struts.xml 中的配置找到对应的返回结果。

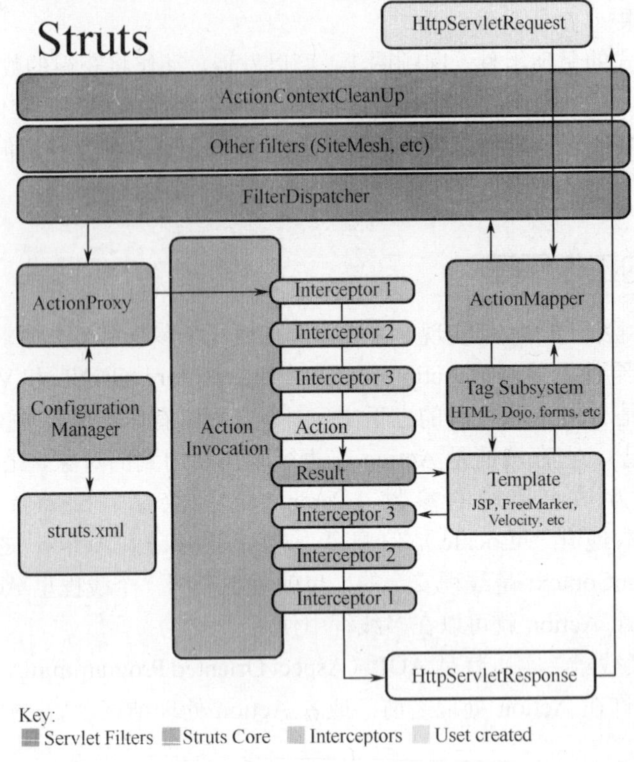

图 1-2　Struts2 的工作原理图

使用 Action 代理正是 Struts2 设计的精妙处。它可以根据系统的配置，加载一系列的拦截器，由拦截器将 HttpServletRequest 参数解析出来，传入 Action。同样，Action 处理的结果也是通过拦截器传入 HttpServletResponse 的，然后由 HttpServletRequest 传给用户。

1.2　Struts2 开发

下面接着讲述 Struts2 的开发，首先介绍 Struts2 Web 应用程序常用的配置文件，接着用一个实例讲解 Struts2 在 Eclipse 中的开发步骤。

Struts2 的配置文件

Struts2 开发过程中主要用到的配置文件包括 web.xml 和 struts.xml，下面将分别介绍这些配置文件。

1. web.xml

web.xml 是在 Servlet 规范中定义的一个部署描述符文件。任何 MVC 框架需要与 Web 应用整合，还得借助于 Web.xml 文件实现，这样配置在 web.xml 文件中 Servlet 才会被应用加载。

通常，所有的 MVC 框架都需要 Web 应用加载一个核心控制器，对于 Struts2 框架而言，需要加载 FilterDispatcher，只要 Web 应用负责加载 FilterDispatcher，FilterDispatcher 将会加载应用的 Struts2 框架。

Struts2 将核心控制器设计成 Filter，而不是一个普通 Servlet，是为了让 Web 应用加载 FilterDispacher，只需要在 web.xml 文件中配置 FilterDispatder 即可。Struts2 中 FilterDispacher 的典型配置如下：

```
<filter>
        <filter-name>struts2</filter-name>
        <filter-class>
org.apache.struts2.dispatcher.ng.filter.StrutsPrepareAndExecuteFilter
</filter-class>
    </filter>
    <filter-mapping>
        <filter-name>struts2</filter-name>
        <url-pattern>/*</url-pattern>
    </filter-mapping>
```

2. struts.xml

struts.xml 作为开发中使用最为频繁的一个重要的配置文件，它的结构包括该文件的根元素及每个元素能包含的子元素等。本节主要讲解包配置、命名空间配置和常量配置等。

包配置

框架会把动作组件和其他的组件一起放在一种叫做包（package）的逻辑容器内。Struts2 的包很像 Java 包，它提供了一种基于功能或者领域的共性将动作组件分组的机制。一些重要的操作属性，例如用来映射到动作的 URL 命名空间，都是在包级别定义的。另外包还提供了一种继承机制，除了其他的特性之外还能够继承框架已经定义的组件。在包上只能设置 4 个属性：name、namespace、extends 和 abstract。表 1-1 总结了这些属性。

表 1-1 Struts2 包的属性

属 性	描 述
name	必需，包的名字
namespace	包内所有动作的命名空间
extends	被继承的父包
abstract	如果为 true，这个包只能用来定义可继承的组件，不能定义动作

name 是必需的属性，name 属性只是一个逻辑名，通过它可以引用这个包。

namespace 属性用来生成这些包内动作被映射到的 URL 命名空间。如果不设置 namespace 属性，动作就会进入默认命名空间。

使用 struts-default 包中定义的智能默认组件很容易，只需要在创建自己的包时继承这个默认包即可。根据定义，智能默认值不需要开发人员手动做任何操作。一旦你继承了这个包，很多组件就自动开始起作用了。

命名空间

Struts2 通过为包指定 namespace 属性来为包下面的所有 Action 指定共同命名空间。包和命名空间的常见配置代码如下：

```
<package name="default" namespace="/" extends="struts-default">
</package>
```

常量配置

通过配置常量，可以改变Struts2框架和插件的行为，从而满足不同Web应用的需求。实际上，配置常量就是配置Struts2的属性。常量可以在多个文件中声明。Struts2框架默认按照下列文件的顺序搜索常量，顺序靠后的文件中的常量设置可以覆盖前面文件中的常量设置。文件加载的先后顺序如下：

struts_default.xml→struts_plugin.xml→struts.xml→struts.properties→web.xml

在struts.xml配置文件中，通常使用<constant>标签配置常量。使用<constant>元素配置常量时，需要定义以下两个必填属性。

（1）name：指定常量的名称；

（2）value：指定常量的属性值。

下面是一个典型的常量配置代码：

```
<constant name="struts.custom.i18n.resources" value="message"/>
<constant name="struts.i18n.encoding" value="utf-8"></constant>
```

包含配置

在大型的Web项目中，为了降低项目的复杂度，便于团队成员分工合作，同时也是为了提高项目开发效率，通常将项目划分为多个小模块，每个模块单独开发与管理，Struts2也提供了这种功能。实践中，Struts2可以为每个小模块提供一个配置文件，对其进行配置，然后在struts.xml文件中使用include元素来包含其他的配置文件。实现的关键代码如下：

```
<?xml version="1.0" encoding="UTF-8" ?>
<!DOCTYPE struts PUBLIC
        "-//Apache Software Foundation//DTD Struts Configuration 2.3//EN"
        "http://struts.apache.org/dtds/struts-2.3.dtd">
<struts>
      <include file="test.xml"></include>
</struts>
```

1.3 Struts2开发入门

在创建第一个Dynamic Web Project之前，需要安装JDK以及设置Java开发环境变量，JDK的安装以及开发环境变量的设置参看附录A。前面介绍了Struts2框架的运行以及所需的配置文件，现在开始搭建Struts2的第一个案例以及测试案例。

1.3.1 创建工程

（1）首先进入Apache的官方主页http://struts.apache.org，单击Download链接，下载Struts2的最新稳定版，目前Struts2的最新版为2.3.20，如图1-3所示。然后选择Full Distribution下载，将其下载到电脑，解压（提取）到如图1-4所示的目录中。

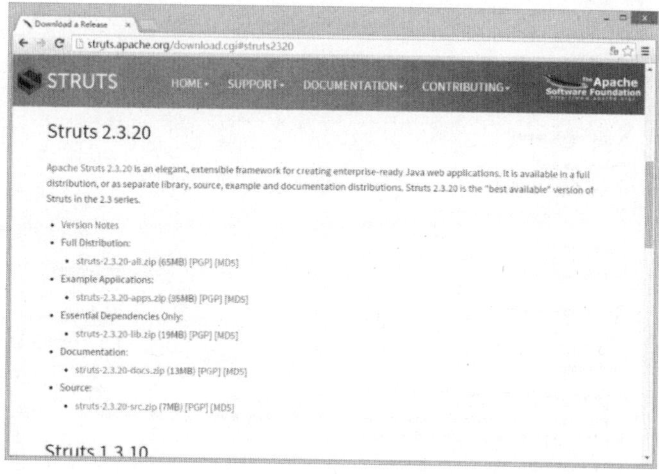

图 1-3　下载 Struts2 开发包

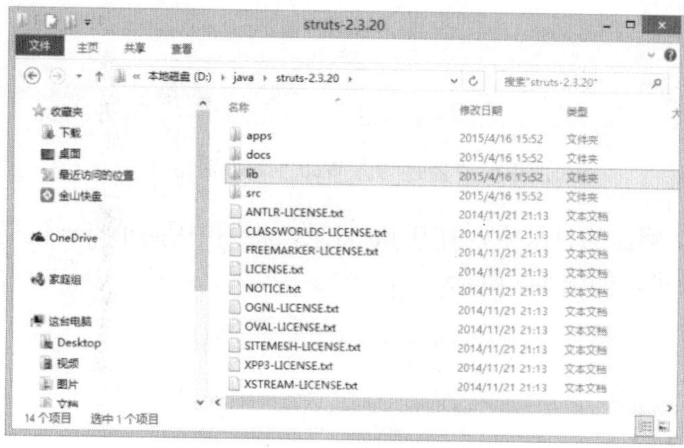

图 1-4　Struts2 开发包目录结构

（2）打开 Eclipse 新建 Web 工程，执行“File→New→Project”菜单命令，出现如图 1-5 所示对话框。选择 Web\Dynamic Web Project，单击“Next”按钮。接着创建如图 1-6 所示的工程，工程名为 struts1。

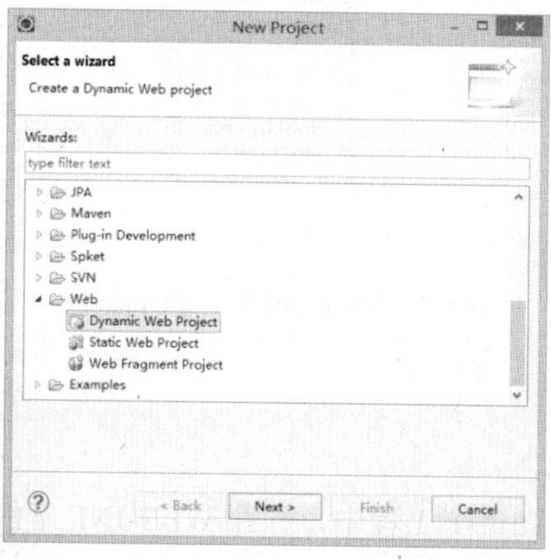

图 1-5　“新建工程”对话框

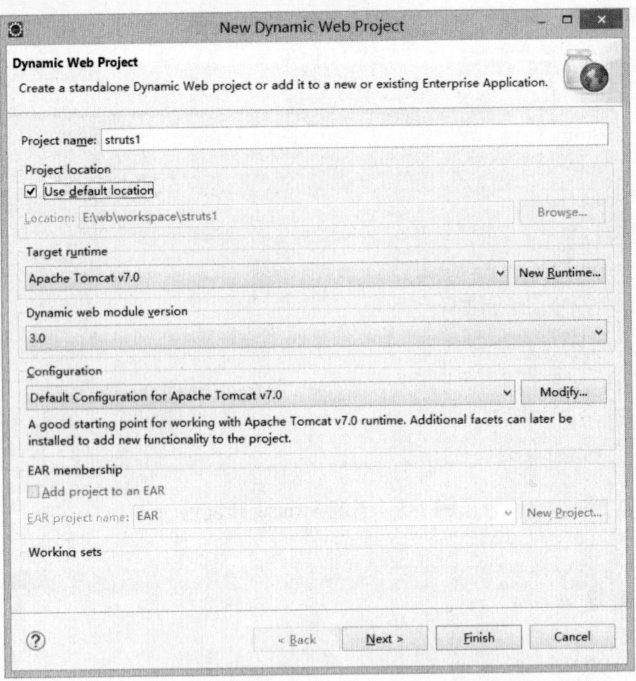

图 1-6 “新建动态 Web 工程”对话框

（3）进入 Struts2 解压的目录中，打开 lib 目录，将图中所示的 jar 包复制到工程 struts1 中的 WEB-INF\lib 目录中，如图 1-7 所示。

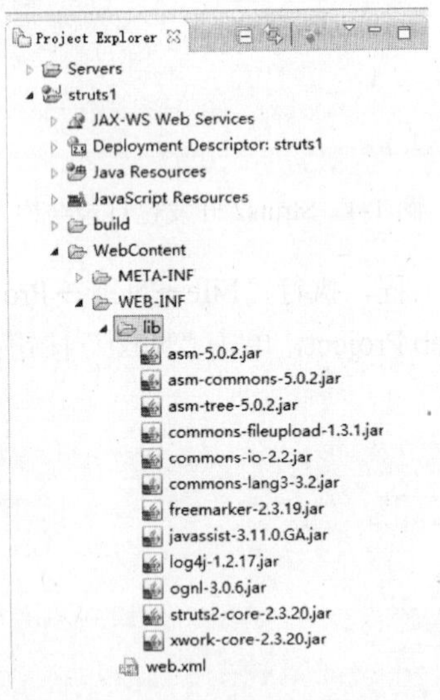

图 1-7　Struts2 开发需要的 jar 包

1.3.2　编写配置文件

（1）如图 1-10 所示，在 struts1 工程中，打开 WEB-INF 下的 web.xml 文件，添加下面的代码：

```
        <filter>
            <filter-name>struts2</filter-name>
            <filter-class>
    org.apache.struts2.dispatcher.ng.filter.StrutsPrepareAndExecuteFilter
    </filter-class>
        </filter>
        <filter-mapping>
            <filter-name>struts2</filter-name>
            <url-pattern>/*</url-pattern>
        </filter-mapping>
```

（2）实现在 src 中新建 struts.xml 文件。首先右击 struts1 工程下的 Java Resources→src，然后选择“Other”，出现新建对话框，如图 1-18 所示，单击“Next”按钮，出现新建 XML 文件对话框，如图 1-9 所示。

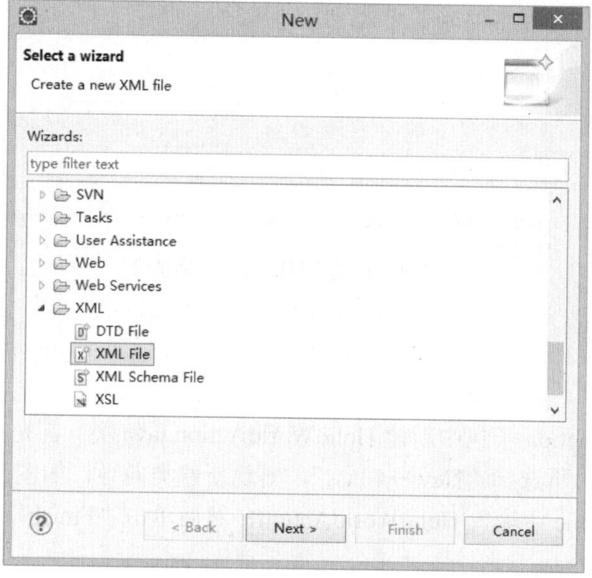

图 1-8　新建 XML 文件对话框 1

（3）在 File name 中键入 struts.xml，单击“Finish”按钮，然后将 struts.xml 的内容修改为：

```
<?xml version="1.0" encoding="UTF-8" ?>
<!DOCTYPE struts PUBLIC
        "-//Apache Software Foundation//DTD Struts Configuration 2.3//EN"
        "http://struts.apache.org/dtds/struts-2.3.dtd">
<struts>
    <package name="default" namespace="/" extends="struts-default">

    </package>
</struts>
```

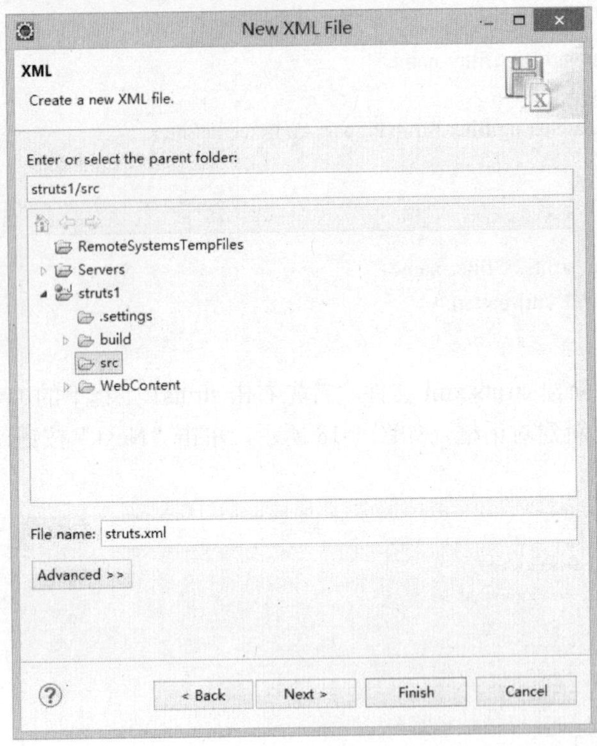

图 1-9　新建 XML 文件对话框 2

1.3.3　编写源码

（1）在 src 的 cap.action 子包中新建 HelloWorldAction.java 类。首先右击“Java Resources→src”，如图 1-10 所示，再选择“New→Class”，出现新建类向导，如图 1-11 所示。在 Package 中输入 cap.action，Name 中输入 HelloWorldAction，然后单击“Finish”按钮。

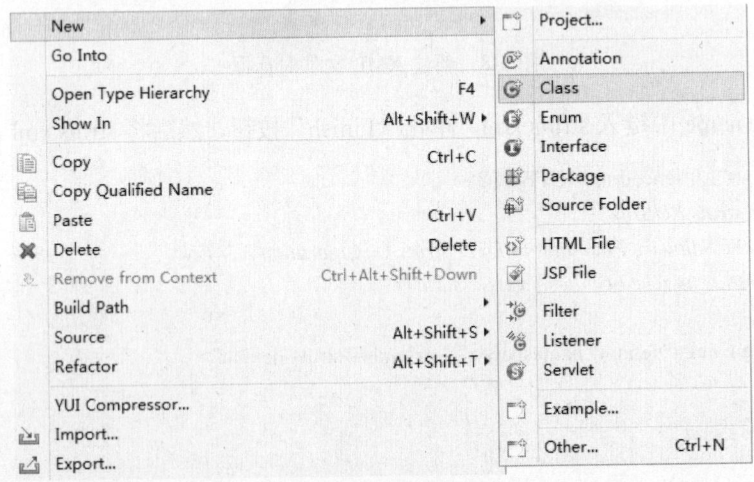

图 1-10　新建类 1

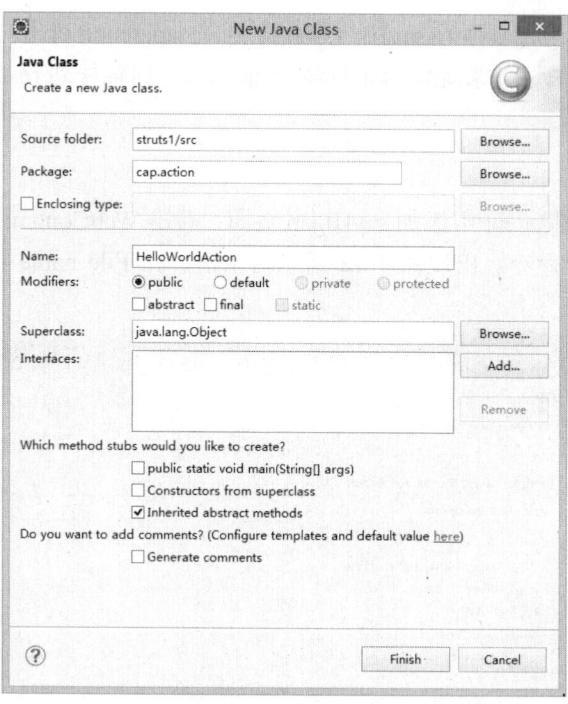

图 1-11　新建类 2

（2）修改 HelloWorldAction.java 的内容。首先 HelloWorldAction 要继承 ActionSupport，编辑后的代码如下：

```
package cap.action;
import com.opensymphony.xwork2.ActionSupport;
@SuppressWarnings("serial")
public class HelloWorldAction extends ActionSupport{
        private String username;
        public String getUsername() {
                return username;
        }
        public void setUsername(String username) {
                this.username = username;
        }
        public String execute(){
                return SUCCESS;
        }
}
```

代码解释：在 Struts2 中定义动作默认调用的方法为 execute()方法，在方法体只是返回了一个 SUCCESS 常量，用于在 Struts2 的配置文件 strtuts.xml 中定位视图。

（3）在 struts.xml 中添加 action 映射(mapping)，代码如下：

```
<action name="sayHi" class="cap.action.HelloWorldAction">
        <result name="success">/result.jsp</result>
</action>
```

代码解释：定义动作的名称为 sayHi，对应的类为 cap.action 子包下的 HelloWorldAction，并且定义了一个返回结果，如果返回的字符串为 success，则跳转到 result.jsp 页面。

1.3.4 编写视图

（1）在工程的 WebContent 中添加 sayHi.jsp 页面：右击 WebContent，在弹出的快捷菜单中选择“New→JSP File”命令。出现图 1-12 所示对话框，在 File name 中输入 sayHi.jsp，单击“Next”按钮。

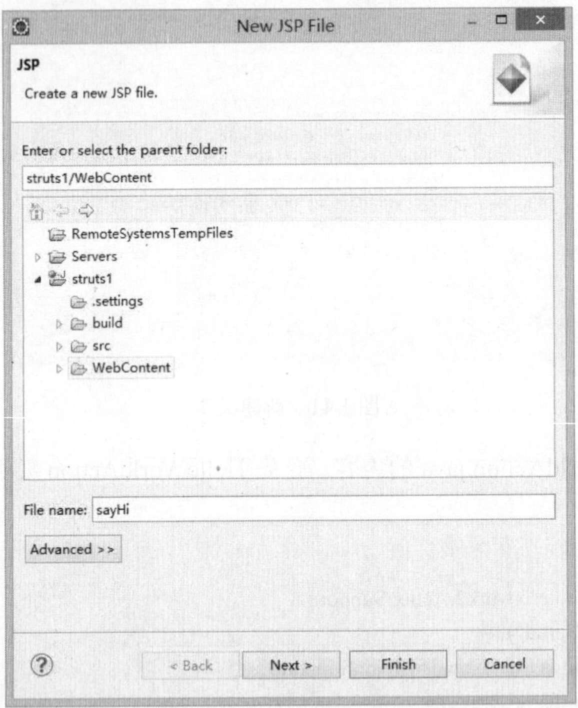

图 1-12 新建 JSP File 对话框

（2）修改 sayHi.jsp 的内容，在 body 标签中添加下面的代码：

```
<form action="sayHi" method="post">
    <input type="text" name="username">
    <input type="submit" value="提交">
</form>
```

（3）新建 result.jsp：修改后的代码如下：

```
<%@ page language="java" contentType="text/html; charset=UTF-8"
    pageEncoding="UTF-8"%>
<%@ taglib uri="/struts-tags" prefix="s"%><!--①-->
<!DOCTYPE html PUBLIC "-//W3C//DTD HTML 4.01 Transitional//EN" "http://www.w3.org/TR/html4/loose.
dtd">
<html>
<head>
<meta http-equiv="Content-Type" content="text/html; charset=UTF-8">
```

```
<title>result</title>
</head>
<body>
<s:property value="username"/><!--② -->
</body>
</html>
```

代码解释：其中①代码用于引入 Struts2 的标签库，标签库的具体使用会在第 3 章详细讲述，②代码将用户输入的用户名显示到页面中。

1.3.5 运行应用程序

（1）启动 tomcat 运行测试，打开浏览器，键入网址 http://localhost:8080/struts1/sayHi.jsp，如图 1-13 所示。在数据框中输入“成都航空职业技术学院”，单击“提交”按钮之后运行的结果如图 1-14 所示。

图 1-13 工程 struts1 运行结果 1

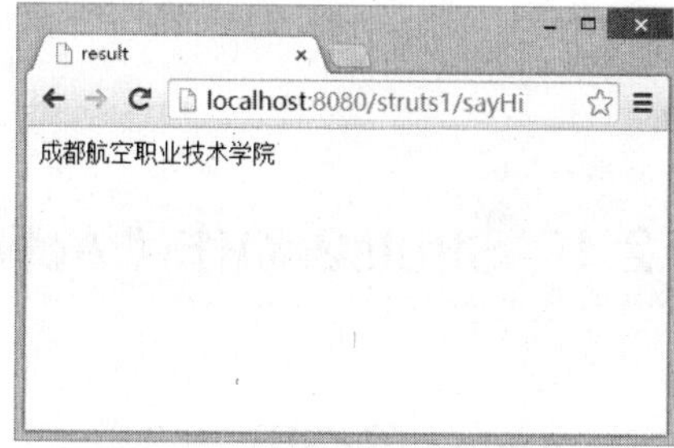

图 1-14 工程 struts1 运行结果 2

第2章 Struts2核心组件

2.1 Struts2动作（Action）

2.1.1 Action的作用

Struts2的动作（Action）组件是Struts2框架的核心。Action主要有三个作用：第一，为给定的请求封装需要做的实际工作；第二，在从请求到视图的框架自动数据传输中作为中介；第三，帮助框架确定要返回的视图。

1. 动作封装工作单元

动作在框架中可作为MVC模式的模型。这个角色的主要职责是控制业务逻辑，动作使用execute()方法来实现业务逻辑。这个方法中的代码应该只关注与请求相关的工作逻辑。

2. 动作为数据转移提供场所

动作是框架的模型组件，使得动作能够携带数据。由于数据保存在动作本地，在执行业务逻辑的过程中就可以很方便地访问到它们。或许一系列的JavaBean属性会增加动作组件的代码量，但是当execute()方法引用这些属性中存储的数据时，邻近的数据会让动作的代码变得更简洁。

3. 动作为结果路由选择返回控制字符串

动作组件的最后一个职责是返回控制字符串以选择应该被呈现的结果页面。先前的框架将路由对象（ActionMapping）整体传递给了动作组件的入口方法。返回的控制字符串消除了对这些对象的依赖，使得方法签名更简洁，并且降低了动作与具体路由代码的耦合。返回字符串的值必须与配置在声明性架构中期望的结果的名字匹配。例如，HelloWorldAction返回“SUCCESS”字符串，可以从XML配置文件中看出，SUCCESS是某一个结果组件的名字。

2.1.2 Action 类的编写

为了让用户开发的 Action 类更加规范，Struts2 提供了一个 Action 接口，这个接口定义了 Struts2 的 Action 类应该实现的规范。下面是标准 Action 接口的代码：

```
public interface Action {
    //定义 Action 接口里包含的一些结果字符串
    public static final String ERROR = "error";
    public static final String INPUT = "input";
    public static final String LOGIN = "login";
    public static final String NONE = "none";
    public static final String SUCCESS = "success";
    //定义处理用户请求的 execute()方法
    public String execute() throws Exception;
}
```

上面的 Action 接口里只定义了一个 execute()方法，该接口规范规定了 Action 类应该包含一个 execute()方法，该方法返回一个字符串，此外，该接口还定义了 5 个字符串常量，它的作用是统一 execute()方法的返回值。例如，当 Action 类处理完用户请求后，有人喜欢返回 welcome 字符串，有人喜欢返回 success 字符串，如此不利于项目的统一管理，Struts2 的 Action 接口定义加上了如上的 5 个字符串常量：ERROR、NONE、INPUT、LOGIN、SUCCESS 等，分别代表了特定的含义。当然，如果开发者依然希望使用特定的字符串作为逻辑视图名，开发者依然可以返回自己的视图名。

另外，Struts2 还为 Action 接口提供了一个实现类：ActionSupport，下面是该实现类部分的代码，具体的实现代码读者可以参考 Struts2 的源码包。

```
public class ActionSupport implements Action, Validateable, ValidationAware,
            TextProvider, LocaleProvider, Serializable {
    // 默认处理用户请求的方法，直接返回 SUCCESS 字符串
    public String execute() throws Exception {
            return SUCCESS;
    }
    @Override
    public Locale getLocale() {
            return null;
    }
    @Override
    public String getText(String arg0) {
            return null;
    }
    @Override
    public ResourceBundle getTexts() {
            return null;
    }
    @Override
    public void addActionError(String arg0) {}
    @Override
```

```
    public void validate() { }
    //省略部分需要实现接口的方法
}
```

ActionSupport 是一个默认的 Action 实现类，该类里已经提供了许多默认方法，这些方法包括获取国际化信息的方法、数据校验的方法、默认的处理用户请求的方法等。实际上，ActionSupport 是 Struts2 默认的 Action 处理类，如果让开发者的 Action 类继承该 ActionSupport 类，则会大大简化 Action 的开发。

2.1.3 Action 的使用与配置

现在可以开始开发动作了。在本节中，首先讲解 Action 的配置，接着讲解 Action 类的具体编写使用。Struts2 的 Action 的配置是在 struts.xml 中进行的，通过<action>元素进行配置，<action>元素常用的属性有如下几个：

- name：该属性用来指定客户端发送请求的地址映射名称；
- class：该属性用来指定业务逻辑处理的 Action 类名称；
- method：该属性用来指定进行业务逻辑处理的 Action 类中的方法名称。

一个典型的 action 配置如下：

```
<action name="sayHello" class="cap.action.HelloWorldAction" method="sayHello">
  <result name="success">/output.jsp</result>
</action>
```

在上述 Action 代码配置中，为 action 指定 name，class 和 method 属性。method 属性值就是 Action 类中定义的方法名，在默认情况下是 execute()方法。

通常要为 Action 指定一个或多个视图，这些视图的名称或类型通过 result 元素来配置，该元素主要有 type 和 name 属性。result 元素的 name 属性用来指定 Action 中方法返回的名称，例如上面的代码片段返回 SUCCESS。

1. 动态 Action 调用

实际应用中，每个 Action 都要处理很多业务，所以每个 Action 都要包含多个处理业务逻辑的方法。针对不同的客户端请求，Action 都会调用不同的方法来处理。Struts2 提供使用 method 属性实现方法的动态调用。实现的步骤如下：

首先在表单中的 action 中指定 action 的名称，例如下面的代码：

```
<form action="login" method="post">
```

其次在 struts.xml 中配置 action 名称：

```
<action name="login" class="cap.action.LoginAction" method="login">
    <result name="success">/output.jsp</result>
</action>
```

下面将通过一个具体的例子讲解动态 Action 调用的方法。

（1）继续在 struts1 工程中，在 HelloWorldAction 中添加 SayHello 的方法，在 struts2 中定义方法的原型如下：

```
public String someMethod(){}
```

（2）在 src 下的 cap.action 子包中的 HelloWorldAction.java 文件中添加 SayHello 方法，实现代码如下：

```
public String sayHello(){
        username="欢迎您：" +username;
        return SUCCESS;
    }
```

（3）在 src 中的 sturts.xml 文件中写入以下的代码：

```
<action name="sayHello" class="cap.action.HelloWorldAction" method="sayHello">
    <result name="success">/output.jsp</result>
</action>
```

（4）在 WebContent 下复制 sayHi.jsp，重命名为 sayHello.jsp。将 form 标签中 action 属性的值修改为 sayHello。

（5）运行工程：在浏览器中输入 http://localhost:8080/struts1/sayHello.action 地址来访问 HelloWorld Action。如图 2-1 所示，输入“starlee2008@163.com”文本，单击“提交”按钮。返回结果页面如图 2-2 所示。

图 2-1 sayHello 运行结果 1

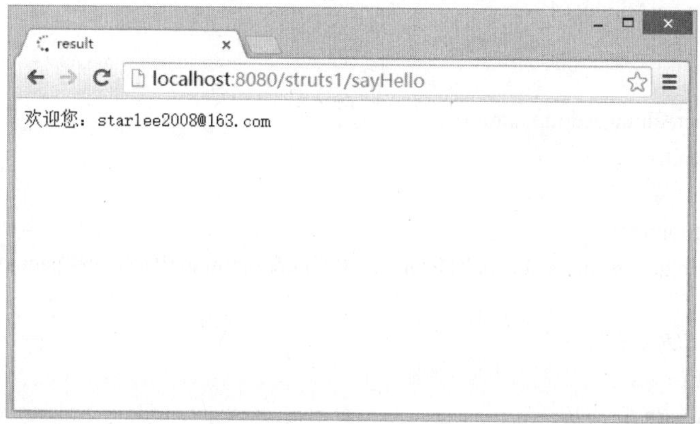

图 2-2 sayHello 运行结果 2

通过上面的例子，可以发现 src/cap.action 子包中 HelloWorldAction.java 类的 Action 方法

（execute 和 sayHello）返回的都是 SUCCESS。这个属性变量在 ActionSupport 类或其父类中定义。

2. 向对象传递数据

使用 Struts2，用对象传递数据将变得非常方便，和普通的 POJO（Plain Ordinary Java Object，简单的 Java 对象，实际就是普通 JavaBeans，是为了避免和 EJB 混淆所创造的简称词）一样在 Action 编写 Getter 和 Setter，然后在 JSP 的 UI 标志的 name 与其对应，在提交表单到 Action 时，就可以获得其值。

下面我们看一个具体向对象传递数据的例子。

（1）按照第 1 章的内容在 Eclipse 中新建 Dynamic Web Project 工程 struts2。首先在 WebContent 下新建 login.jsp 页面，在 body 标签中添加下面的代码：

```
<form action="login" method="post">
  <input type="text" name="admin.username"><br>
  <input type="password" name="admin.password"><br>
  <input type="submit" value="登录">
</form>
```

（2）在 src 的 cap.bean 包中新建 Admin.java 类，编辑后的代码如下：

```
package cap.bean;
public class Admin {
	private int id;
	private String username;
	private String password;
	//省略 getters 和 setters
}
```

（3）在 src 的 cap.action 包中新建 LoginAction 类，编辑后的代码如下：

```
package cap.action;
import cap.bean.Admin;
import com.opensymphony.xwork2.ActionSupport;
public class LoginAction extends ActionSupport{
	private Admin admin;
	public Admin getAdmin() {
		return admin;
	}
	public void setAdmin(Admin admin) {
		this.admin = admin;
	}
	public String login(){
		if(admin.getUsername().equals("starlee2008")&&admin.getPassword().equals("starlee2008"))
		{
		return SUCCESS;
		}else{
			return ERROR;
		}
	}
}
```

代码解释：这里没有采用数据库验证登录的方式，采用的是静态判断，只要输入的用户名和密码都为“starlee2008”时，返回登录成功的SUCCESS字符串，反之返回ERROR的字符串。

（4）在src的struts配置文件struts.xml中添加下面的代码：

```
<action name="login" class="cap.action.LoginAction" method="login">
        <result name="success">/result.jsp</result>
        <result name="error">/error.jsp</result>
</action>
```

（5）还需添加登录判断后的两种跳转页面，分别是result.jsp和error.jsp，可查看随书提供的示例源码。运行Tomcat，在浏览器地址栏中键入http://localhost:8080struts2/login.jsp，出现如图2-3所示页面。在User 和name中分别输入“cap”和“cap”，单击“登录”按钮，出现如图2-4所示页面。如果输入了错误的用户名和密码，会提示登录失败。

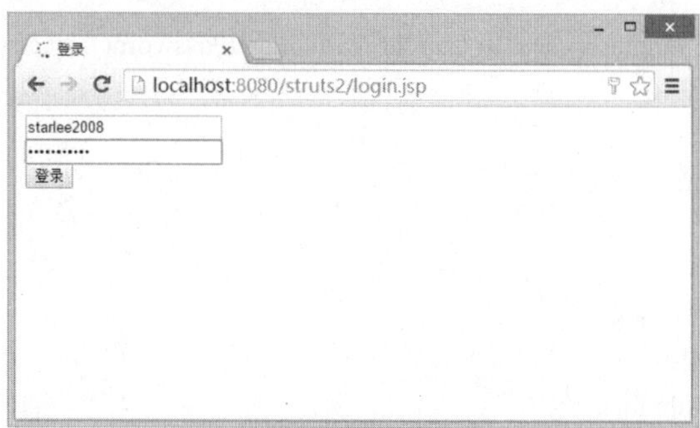

图2-3 login.jsp运行

图2-4 登录结果

3. 模型驱动

所谓模型驱动，就是使用单独的JavaBean实例来贯穿整个MVC流程。简单地说，模型驱动就是使用单独的VO(Value Object，值对象)来封装请求参数和处理结果。

使用模型驱动模式时，Action必须实现ModelDriven接口，实现该接口则必须实现getModel方法，该方法用于把Action和与之对应的Model实例关联起来。

下面继续通过一个案例讲解模型驱动的使用。

（1）继续在工程struts2中操作。在src的cap.action包中新建LoginxAction类，编辑后的代码如下：

```
package cap.action;
import cap.bean.Admin;
import com.opensymphony.xwork2.ActionSupport;
import com.opensymphony.xwork2.ModelDriven;
@SuppressWarnings("serial")
public class LoginxAction extends ActionSupport implements ModelDriven<Admin>{
        private Admin admin;
        public Admin getAdmin() {
                return admin;
        }
        public void setAdmin(Admin admin) {
                this.admin = admin;
        }
        public String loginx(){
                if(admin.getUsername().equals("cap")&&admin.getPassword().equals("cap"))
                return SUCCESS;
                else
                return ERROR;
        }
        @Override
        public Admin getModel() {
                if(admin==null){
                admin=new Admin();
                }
                return admin;
        }
}
```

代码解释: loginx 方法和前一节的方法功能相同，实现静态的用户名和密码判断，getModel 方法主要是创建一个 Admin 的对象 admin，并根据页面 loginx.jsp 传递过来的 username 和 password 值，通过调用对象 admin 的 setUsername 方法和 setPassword 方法设置值。

（2）在 WebContent 中新建 loginx.jsp，在 body 标签中添加如下的代码：

```
<form action="loginx" method="post">
        <input type="text" name="username"><br>
        <input type="password" name="password"><br>
        <input type="submit" value="登录">
</form>
```

代码解释: 和 login.jsp 页面的主要区别是本页面采用传递的是属性值，而不是采用对象名.属性值（admin.username）的方法传递值。实现 ModelDriven 接口后，会通过 getModel 创建 Admin 对象 admin，并初始化其值。

（3）在 src 下的 struts.xml 文件中添加下面的代码：

```
<action name="loginx" class="cap.action.LoginxAction" method="loginx">
    <result name="success">/result.jsp</result>
    <result name="error">/error.jsp</result>
</action>
```

2.2 Struts2 拦截器（Interceptor）

2.2.1 拦截器

拦截器实质就是 AOP(面向切面编程)的编程思想。拦截器允许在 Action 处理之前，或者 Action 处理结束之后，插入开发者自定义的代码。

拦截器体系是 Struts2 的一个重要组成部分，对于 Struts2 框架而言，可以将其理解为一个空容器，正是大量的内建拦截器完成了该框架的大部分操作。

对于 Struts2 的拦截器体系而言，当需要使用某个拦截器时，只需要在配置文件中应用该拦截器即可，如果不需要使用该拦截器，也只需要在配置文件中取消该拦截器，不管是否应用某个拦截器，对于整个 Struts2 框架不会有任何影响。这种设计方式非常灵活，可扩展性好。

因为 Struts2 框架的拦截器是动态配置的，所以开发者可以非常方便地扩展 Struts2 框架，只要实现 Interceptor 接口或者继承 AbstractInterceptor 类，并将其配置在 struts.xml 文件中即可。实际上，在 Struts2 中开发自定义拦截器非常便利，因此，开发者可非常方便地将多个 Action 中需要重复执行的动作放在拦截器中定义，从而提供更好的代码复用。

虽然拦截器很重要，但是一般不会去编写很多拦截器。实际上，Web 应用程序领域常见任务已经编写和捆绑进了 struts-default 包。

1. 拦截器的工作原理

现在来看看拦截器是如何工作的。拦截器 ActionInvocation 类负责指挥着动作的完整执行，以及与之相关的拦截器栈的顺序触发。

ActionInvocation 类封装了与特定动作执行相关的所有处理细节。当框架收到一个请求时，它首先必须决定这个 URL 映射到哪个动作。这个动作的一个实例会被加入到一个新创建的 ActionInvocation 实例中。接着，框架查询声明性架构(通过应用程序的 XML 或者 Java 注解创建)，以发现哪些拦截器应该触发，以及按照什么样的顺序触发。指向这些拦截器的引用被加入到 ActionInvocation 类中。除了这些核心元素，ActionInvocation 还具有其他重要信息。

ActionInvocation 类创建好并且填充了需要的所有对象和信息，就可以开始调用。ActionInvocation 类公开了 invoke()方法，框架通过调用这个方法开始动作的执行。当框架调用这个方法时，ActionInvocation 通过执行拦截器栈中的第一个拦截器开始这个调用过程，但需注意，invoke()方法并不总是映射到第一个拦截器。

如图 2-5 所示，第一个被触发的拦截器是 exception 拦截器。从这里开始，每一个拦截器按照与从上到下相同的顺序触发。因此，最后一个被触发的拦截器会是 workflow 拦截器。在结果被呈现之后，拦截器会按照相反的顺序再触发一遍，使它们有机会做后续处理。

后续拦截器继续执行，最终执行动作，这些都通过递归调用 ActionInvocation 的 invoke()方法实现。每次 invoke()方法被调用时，ActionInvocation 都会查询自身的状态，调用接下来的任何一个拦截器。在所有的拦截器都被调用之后，invoke ()方法会促使动作类执行。

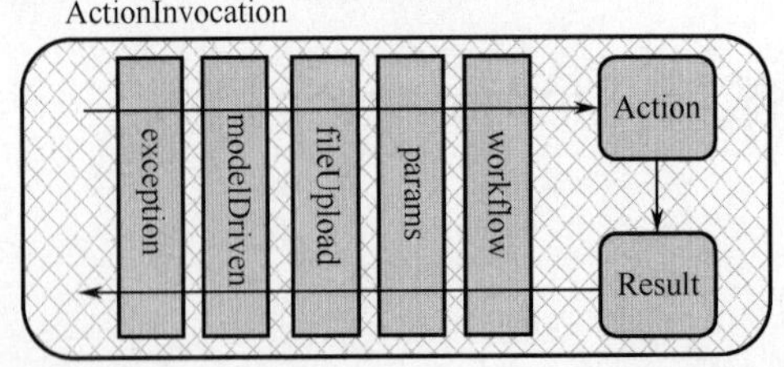

图 2-5　ActionInvocation 封装了动作及与之关联的拦截器和结果的执行

拦截器的执行周期，主要分为三个阶段：

第一阶段，做一些预处理。

第二阶段，通过调用 invoke()方法将控制转移给后续的拦截器，最后直到动作，或者通过返回一个控制字符串中断执行。

第三阶段，做一些后加工。

通常，使用拦截器可以解决如权限控制、跟踪日志、跟踪系统的性能瓶颈等问题。

2. 拦截器的配置

Struts2 框架自带很多内置栈，使内置拦截器方便排列。可以通过扩展 struts-default.xml 文件中定义的 struts-default 包来继承包含 defaultStack 在内的所有内建的栈。拦截器的配置代码片段如下：

```
<action name="timer" class="cap.action.TimerAction">
        <interceptor-ref name="timer"></interceptor-ref>
        <interceptor-ref name="logger"></interceptor-ref>
        <result name="success">/result.jsp</result>
</action>
```

上述代码实现将 timer 和 logger 拦截器与动作（timer）关联起来，它们会按被列出的顺序触发。Struts2 已经提供丰富多样的，功能齐全的拦截器实现。在 struts2-core-2.x.x.jar 包的 struts-default.xml 文件中可以查看关于默认的拦截器与拦截器链的配置。

下面列出了在 struts-default.xml 文件定义的部分拦截器：

```
<interceptors>
        <interceptor name="alias" class="com.opensymphony.xwork2.interceptor.AliasInterceptor"/>
        <interceptor name="autowiring" class="com.opensymphony.xwork2.spring.interceptor.
ActionAutowiringInterceptor"/>
        <interceptor name="chain" class="com.opensymphony.xwork2.interceptor.ChainingInterceptor"/>
        <interceptor name="conversionError" class="org.apache.struts2.interceptor.
StrutsConversionErrorInterceptor"/>
        <interceptor name="cookie" class="org.apache.struts2.interceptor.CookieInterceptor"/>
        <interceptor name="cookieProvider" class="org.apache.struts2.interceptor.
CookieProviderInterceptor"/>
        <interceptor name="clearSession" class="org.apache.struts2.interceptor.
ClearSessionInterceptor" />
        <interceptor name="createSession" class="org.apache.struts2.interceptor.
CreateSessionInterceptor" />
        <interceptor name="debugging" class="org.apache.struts2.interceptor.debugging.
DebuggingInterceptor" />
```

```
                <interceptor name="execAndWait" class="org.apache.struts2.interceptor.
ExecuteAndWaitInterceptor"/>
                <interceptor name="exception" class="com.opensymphony.xwork2.interceptor.
ExceptionMappingInterceptor"/>
                <interceptor name="fileUpload" class="org.apache.struts2.interceptor.FileUploadInterceptor"/>
                <interceptor name="i18n" class="com.opensymphony.xwork2.interceptor.I18nInterceptor"/>
                <interceptor name="logger" class="com.opensymphony.xwork2.interceptor.LoggingInterceptor"/>
                <interceptor name="modelDriven" class="com.opensymphony.xwork2.interceptor.
ModelDrivenInterceptor"/>
                <interceptor name="scopedModelDriven" class="com.opensymphony.xwork2.interceptor.
ScopedModelDrivenInterceptor"/>
                <interceptor name="params" class="com.opensymphony.xwork2.interceptor.
ParametersInterceptor"/>
                <interceptor name="actionMappingParams" class="org.apache.struts2.interceptor.
ActionMappingParametersInteceptor"/>
                <interceptor name="prepare" class="com.opensymphony.xwork2.interceptor.
PrepareInterceptor"/>
                <interceptor name="staticParams" class="com.opensymphony.xwork2.interceptor.
StaticParametersInterceptor"/>
                <interceptor name="scope" class="org.apache.struts2.interceptor.ScopeInterceptor"/>
                <interceptor name="servletConfig" class="org.apache.struts2.interceptor.
ServletConfigInterceptor"/>
                <interceptor name="timer" class="com.opensymphony.xwork2.interceptor.TimerInterceptor"/>
                <interceptor name="token" class="org.apache.struts2.interceptor.TokenInterceptor"/>
                <interceptor name="tokenSession" class="org.apache.struts2.interceptor.
TokenSessionStoreInterceptor"/>
                <interceptor name="validation" class="org.apache.struts2.interceptor.validation.
AnnotationValidationInterceptor"/>
                <interceptor name="workflow" class="com.opensymphony.xwork2.interceptor.
DefaultWorkflowInterceptor"/>
                <interceptor name="store" class="org.apache.struts2.interceptor.MessageStoreInterceptor" />
                <interceptor name="checkbox" class="org.apache.struts2.interceptor.CheckboxInterceptor" />
                <interceptor name="profiling" class="org.apache.struts2.interceptor.
ProfilingActivationInterceptor" />
                <interceptor name="roles" class="org.apache.struts2.interceptor.RolesInterceptor" />
                <interceptor name="annotationWorkflow" class="com.opensymphony.xwork2.interceptor.
annotations.AnnotationWorkflowInterceptor" />
                <interceptor name="multiselect" class="org.apache.struts2.interceptor.MultiselectInterceptor" />
```

如果您想要使用 struts-default.xml 文件中定义的上述拦截器，还需要在应用程序 struts.xml 文件中进行一些设置，后续将用具体案例来讲解。

3. 使用拦截器栈

Struts2 允许将多个拦截器组合在一起，形成一个拦截器栈。对于 Struts2 系统而言，多个拦截器组成的拦截器栈对外也表现为一个拦截器。例如，在 Action 执行前同时做登录检查、安全检查和记录日志，则可以把这三个动作对应的拦截器组成一个拦截器栈。定义拦截器栈使用<intercepter-stack .../>元素，并且需要在其中使用<intercepter-ref.../>元素来定义多个拦截器的引用，即该拦截器栈由多个<intercepter-ref .../>元素指定的拦截器组成。下面是典型的拦截器栈的定义代码片段。

```
<interceptors>
        <!-- 添加登录拦截器 -->
        <interceptor name="checkLogin" class="cap.util.CheckLoginInterceptor"/>
        <!-- 新建一个栈，把登录拦截器和默认的栈放进去 -->
        <interceptor-stack name="mystack">
            <interceptor-ref name="defaultStack"/>
            <interceptor-ref name="checkLogin"/>
        </interceptor-stack>
</interceptors>
```

一旦定义了拦截器和拦截器栈后，就可以使用这个拦截器或拦截器栈来拦截 Action 了，拦截器（包含拦截器栈）的拦截行为将会在 Action 的 execute 方法执行之前被执行。通过<interceptor-ref.../>元素可以在 Action 内使用拦截器。

2.2.2 拦截器的使用

1. 预定义拦截器案例

下面将使用实例讲解拦截器 timer 的使用。

（1）在 Eclipse 中新建工程 struts3，在 src 的 cap.action 包中新建 TimerAction 类，编辑后的代码如下：

```
package cap.action;
import com.opensymphony.xwork2.ActionSupport;
@SuppressWarnings("serial")
public class TimerAction extends ActionSupport{
        public String timer()
        {
                try {
                        Thread.sleep(500);//代码①
                } catch (InterruptedException e) {
                        e.printStackTrace();
                }
                return SUCCESS;
        }
}
```

代码解释：代码①调用线程 Thread 类的静态方法实现睡眠 5 秒钟。

（2）在 src 的 struts.xml 文件中 package 标签处添加如下的代码：

```
<action name="timer" class="cap.action.TimerAction">
        <interceptor-ref name="timer"></interceptor-ref>
        <result name="success">/result.jsp</result>
</action>
```

（3）发布运行应用程序，在浏览器的地址栏键入 http://localhost:8080/struts3/timer，在出现 result.jsp 页面后，查看服务器的后台输出。

```
十一月 12, 2013 3:49:30 下午 com.opensymphony.xwork2.util.logging.jdk.JdkLogger info
信息: Executed action [//timer!timer] took 508 ms.
```

在自己电脑上执行 Timer 拦截器所花的时间，可能与上述时间有些不同，其缘由主要是计算机性能差异所造成的。当第一次加载 Timer 时，需要进行一定的初始工作。当你再次运行，会发现时间接近 500ms。

2. 自定义拦截器案例

对于框架（Framework），可扩展性是必不可少的特性。虽然，Struts2 提供丰富多样的拦截器实现，但我们依然可以创建自定义拦截器，并且也很容易实现。

Struts2 的拦截器直接或间接实现接口 com.opensymphony.xwork2.interceptor.Interceptor，另外，我们也习惯性继承 com.opensymphony.xwork2.interceptor.AbstractInterceptor 类。AbstractInterceptor 是 Interceptor 接口的实现类。

从 com.opensymphony.xwork2.interceptor.Interceptor 接口定义来看，Interceptor 接口只定义了 3 个方法。前 2 个方法是典型的生命周期方法，作用是初始化或者清理资源。真正的业务逻辑发生在 intercept()方法中。可以看到，这个方法被递归的 ActionInvocation.invoke()方法调用。

```
public interface Interceptor extends Serializable{
    public abstract void destory();
    public abstract void init();
    public abstract String intercept(ActionInvocation invocation) throws Exception;
}
```

Intercept（ActionInvocation() invocation）：该方法是用户需要实现的拦截动作。就像 Action 的 execute 方法一样，intercept()方法会返回一个字符串作为逻辑视图。如果该方法直接返回了一个字符串，系统将会跳转到该逻辑视图对应的实际视图资源，不会调用被拦截的 Action。该方法的 ActionInvocation 参数包含了被拦截的 Action 的引用，可以通过调用该参数的 invoke 方法，将控制权转给下一个拦截器，或者转给 Action 的 execute 方法。

下面的例子将展示通过继承 AbstractInterceptor，实现授权拦截器。

3. 自定义拦截器案例

（1）继续在工程 struts3 中编辑，在 src 的 cap.util 包下创建登录拦截器 CheckLoginInterceptor 类，编辑后的代码如下：

```
package cap.util;
import java.util.Map;
import cap.action.LoginAction;
import cap.bean.Admin;
import com.opensymphony.xwork2.Action;
import com.opensymphony.xwork2.ActionInvocation;
import com.opensymphony.xwork2.interceptor.AbstractInterceptor;
@SuppressWarnings("serial")
public class CheckLoginInterceptor extends AbstractInterceptor{
    @Override
    public String intercept(ActionInvocation ai) throws Exception {
        System.out.println("开始拦截器拦截");
        Object obj=ai.getAction();
        if(obj instanceof LoginAction){
```

```
            System.out.println("登录 action 不需要拦截");
            return ai.invoke();
        }
        Map<String,Object> session=ai.getInvocationContext().getSession();
        Admin admin=(Admin)session.get("admin");
        if(admin!=null){
            System.out.println("已经登录。不需要拦截");
            return ai.invoke();
        }else{
            System.out.println("你还没有登录。跳转到登录页面");
            return Action.LOGIN;
        }
    }
}
```

代码解释：在 intercept()方法中，首先通过 ActionInvocation 的 getAction 获得对象 object，接着判断对象 object 是否为 LoginAction 的实例，如果是，不需要拦截，并返回。如果不是 LoginAction 对象的实例，首先获得 session 对象，通过检查 session 是否存在键为“admin”的对象 admin，判断用户是否登录。如果用户已经登录，将角色放到 Action 中，调用 Action；否则，拦截直接返回 Action.LOGIN 字符串并调转到相应的视图页面。

（2）在 src 的 cap.action 中创建 LoginAction 类，编辑后的代码如下：

```
package cap.action;
import java.util.Map;
import org.apache.struts2.interceptor.SessionAware;
import com.opensymphony.xwork2.ActionSupport;
import cap.bean.Admin;
public class LoginAction extends ActionSupport implements SessionAware{//代码①
    private Admin admin;
    private Map<String,Object> session;
    public Admin getAdmin() {
        return admin;
    }
    public void setAdmin(Admin admin) {
        this.admin = admin;
    }
    public String login(){
        if(admin.getUsername().equals("cap")&&admin.getPassword().equals("cap")){
            session.put("admin", admin);
            return SUCCESS;
        }else
            return ERROR;
    }

    @Override
    public void setSession(Map<String, Object> session) {
        this.session=session;
```

```
    }
}
```

代码解释：要在 Action 类中使用 Map<String,Object> session 对象，需要实现 SessionAware 接口，login()方法首先判断用户名和密码是否匹配，如果匹配的话使用 session 将 Admin 的对象 admin 存放在 Map 中名为 admin。如果不匹配，返回 ERROR 常量。

（3）在 src 的 cap.bean 中创建 Admin 类，编辑后的代码如下：

```
package cap.bean;
public class Admin {
        private Integer id;
        private String username;
        private String password;
       //省略 getters 和 setters
}
```

（4）修改 src 下的 struts2 的配置文件 struts.xml，在 struts.xml 中包含 test.xml，代码如下：

```
<?xml version="1.0" encoding="UTF-8" ?>
<!DOCTYPE struts PUBLIC
"-//Apache Software Foundation//DTD Struts Configuration 2.3//EN"
"http://struts.apache.org/dtds/struts-2.3.dtd">
<struts>
<package name="login" extends="struts-default">
        <!-- 添加拦截器 -->
        <interceptors>
            <!-- 添加登录拦截器 -->
            <interceptor name="checkLogin" class="cap.util.CheckLoginInterceptor"/>
            <!-- 新建一个栈，把登录拦截器和默认的栈添加进去 -->
            <interceptor-stack name="mystack">
                <interceptor-ref name="defaultStack"/>
                <interceptor-ref name="checkLogin"/>
            </interceptor-stack>
        </interceptors>
        <!-- 修改默认拦截器 -->
        <default-interceptor-ref name="mystack"/>
        <!-- 将 result 设置为全局的，这样就不用在每一个 package 里面都添加拦截器了 -->
        <global-results>
            <result name="login">/login.jsp</result>
        </global-results>
    </package>
    <include file="test.xml"></include>
</struts>
```

代码解释：<package>之间的代码定义了一个包，其名称为 login，namespace 默认值为“/”，并且继承了 Struts2 的 struts-default 包。接着定义了拦截器栈，并将拦截器栈的名称命名为 mystack，其中包含了自定义的拦截器 checkLogin，<default-interceptor-ref>标签把 mystack 拦截器栈设置为默认的拦截器栈，<global-results>标签定义了一个全局的返回值，如果返回的字符串为 LOGIN，那么将返回到 loing.jsp 视图。<include>标签表示包含一个名为 test.xml 的配置文件。

（5）在 src 下创建 test.xml，编辑后的代码如下：

```
<?xml version="1.0" encoding="UTF-8" ?>
<!DOCTYPE struts PUBLIC
"-//Apache Software Foundation//DTD Struts Configuration 2.3//EN"
"http://struts.apache.org/dtds/struts-2.3.dtd">
<struts>
    <package name="default" namespace="/" extends="login">
        <action name="timer" class="cap.action.TimerAction" method="timer">
            <result name="success">/result.jsp</result>
        </action>
        <action name="login" class="cap.action.LoginAction" method="login">
            <result name="success">/index.jsp</result>
            <result name="error">/error.jsp</result>
        </action>
    </package>
</struts>
```

代码解释：这里重新定义了一个 Struts2 的配置文件，包的名称为 default，并且继承了 login 包，这样 login 包里的全局拦截器就会在此包中起到拦截的作用。

（6）在 WebContent 下创建 login.jsp 页面， 在 body 标签中添加下面的代码：

```
<form action="login" method="post">
    <input type="text" name="admin.username"><br>
    <input type="password" name="admin.password"><br>
    <input type="submit" value="登录">
</form>
```

（7）在 WebContent 下创建 index.jsp 页面，在 body 标签中添加下面的代码（效果见图 2-6）：

```
<p>全局拦截器，只有登录后才能使得 timer 生效</p>
<a href="timer">使用内置的 timer 拦截器，会在控制台下打印出 action 执行的时间</a>
```

（8）发布运行应用程序，在浏览器地址栏中输入：http://localhost:8080/struts3/timer 。单击页面中的超链接，由于没有登录，所以会跳转到 login.jsp 页面，如图 2-7 所示。

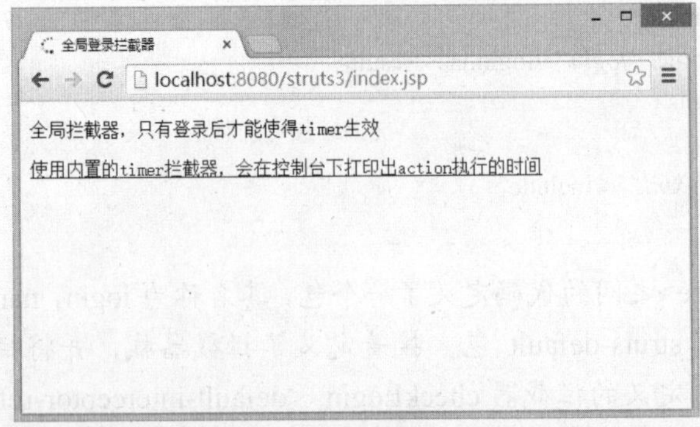

图 2-6　index.jsp 运行

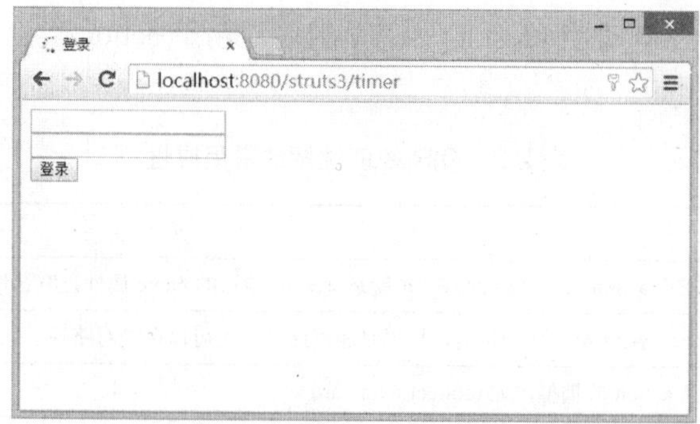

图 2-7　未登录被拦截的结果

2.3　Struts2 注解（Annotation）

注解（Annotation）是 Java 语言比较新的一个特性，它允许在 Java 源代码文件中直接添加元数据。

通常情况下，Struts2 是通过 struts.xml 文件配置的。但是随着系统规模的扩大，所需要配置的文件就会比较多，虽然可以根据不同的系统功能将不同模块的配置文件单独书写，然后通过<include>节点将不同的配置文件引入最终的 struts.xml 文件中，但是毕竟还是要维护和管理这些文件，使得维护工作变得艰巨。

Struts2 声明性架构机制可以配置用来扫描 Java 类以获得与 Struts2 相关的注解。

如果要使用 struts2 的注解功能，必须使用一个包 struts2-convention-plugin-2.x.x.x.jar，其中 xxx 代表版本号。

2.3.1　常用注解

Struts2 中注解的主要概念包括 package、action 以及 Interceptor 等，下面将分别讲解这些概念：

（1）@ParentPackage：此注解对应 xml 文件中的 package 节点，它只有一个属性叫 value，其实就是 package 的 name 属性；

（2）@Namespace：命名空间，也就是 xml 文件中<package>的 namespace 属性；

（3）@Action：此注解对应 <action> 节点，可以应用于 action 类上，也可以应用于方法上。此注解中的常用属性见表 2-1。

表 2-1　@Action 注解的常用属性

属　性	含　义
value	表示 action 的 URL，也就是<action>节点中的 name 属性
results	表示 action 的多个 result，这个属性是一个数组属性，因此可以定义多个 result
interceptorRefs	表示 action 的多个拦截器。这个属性也是一个数组属性，因此可以定义多个拦截器
params	这是一个 String 类型的数组，它按照 name/value 的形式组织，是传给 action 的参数
exceptionMappings	这是异常属性，它是一个 ExceptionMapping 的数组属性，表示 action 的异常，在使用时必须引用相应的拦截器

（4）@Result：这个注解对应<result>节点，只能应用于 action 类上。这个注解中常用属性见表 2-2。

表 2-2　@Result 注解的常用属性

属　性	含　义
name	表示 action 方法的返回值，也就是<result>节点的 name 属性，默认情况下是 success
location	表示 view 层文件的位置，可以是相对路径，也可以是绝对路径
type	是 action 的类型，如 redirect，forward
params	是一个 String 数组，也是以 name/value 形式传送给 result 的参数

2.3.2　注解的使用

下面将通过一个具体的案例来讲解 Struts2 中注解（Annotation）的使用。

（1）在 Eclipse 中，复制 struts3 工程，将工程重新命名为 struts4，如图 2-8 所示。

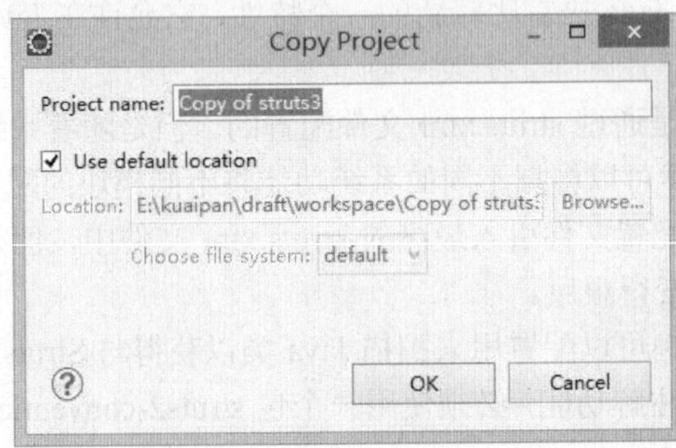

图 2-8　工程复制

（2）将 src 目录下的 test.xml 配置文件删除，然后将 struts.xml 文件中的下面代码删除。

```
<include file="test.xml"></include>
```

（3）编辑 cap.action 下的 LoginActon 类，编辑后的代码如下：

```
package cap.action;
import java.util.Map;
import org.apache.struts2.convention.annotation.Action;
import org.apache.struts2.convention.annotation.InterceptorRef;
import org.apache.struts2.convention.annotation.InterceptorRefs;
import org.apache.struts2.convention.annotation.ParentPackage;
import org.apache.struts2.convention.annotation.Result;
import org.apache.struts2.convention.annotation.Results;
import org.apache.struts2.interceptor.SessionAware;
import com.opensymphony.xwork2.ActionSupport;
import cap.bean.Admin;
@SuppressWarnings("serial")
@ParentPackage("login")
@InterceptorRefs(@InterceptorRef("mystack"))
@Results( { @Result(name = "success", location = "/index.jsp"),
```

```
        @Result(name = "input", location = "/login.jsp") })
public class LoginAction extends ActionSupport implements SessionAware{
    private Admin admin;
    private Map<String,Object> session;
    public Admin getAdmin() {
        return admin;
    }
    public void setAdmin(Admin admin) {
        this.admin = admin;
    }
    @Action(value="login",results={@Result(name="success",location="/index.jsp"),
            @Result(name="error",location="/error.jsp"),
            @Result(name="input",location="/login.jsp")})
    public String login(){
        if(admin.getUsername().equals("cap")&&admin.getPassword().equals("cap")){
            session.put("admin", admin);
            return SUCCESS;
        }else
            return ERROR;
    }
    @Override
    public void setSession(Map<String, Object> session) {
        this.session=session;
    }
}
```

代码解释：注释的具体使用详见表 2-1 和表 2-2。其余的 Java 实现代码和上一节功能相同。

（4）编辑 cap.action 下的 TimerActon 类，编辑后的代码如下：

```
package cap.action;
import org.apache.struts2.convention.annotation.Action;
import org.apache.struts2.convention.annotation.InterceptorRef;
import org.apache.struts2.convention.annotation.InterceptorRefs;
import org.apache.struts2.convention.annotation.ParentPackage;
import org.apache.struts2.convention.annotation.Result;
import org.apache.struts2.convention.annotation.Results;
import com.opensymphony.xwork2.ActionSupport;
@SuppressWarnings("serial")
@ParentPackage("login")
@InterceptorRefs(@InterceptorRef("mystack"))
@Results( { @Result(name = "success", location = "/index.jsp"),
        @Result(name = "input", location = "/login.jsp") })
public class TimerAction extends ActionSupport{
    @Action(value="timer",results={@Result(name="success",location="/result.jsp")})
    public String timer()
    {
        try {
            Thread.sleep(500);
```

```
            } catch (InterruptedException e) {
                e.printStackTrace();
            }
            return SUCCESS;
        }
    }
```

（5）在 WebContent 下的视图页面不变，运行 index.jsp 页面。如果没有登录访问 timer Action，会被拦截并重定向到 login.jsp，实现的效果和 2.2 节（自定义拦截器使用案例）相似，请读者自行验证。

2.4 Struts2 对象图导航语言（OGNL）

2.4.1 OGNL

OGNL 中文全称为对象图导航语言，是 Object-Graph Navigation Language 的缩写。OGNL 是一种强大的技术，语法简单一致，可以存取对象的任意属性，调用对象的方法，它被集成在 Struts2 框架中用来帮助实现数据转移和类型转换等功能。OGNL 在框架中，像是基于字符串的 HTTP 输入/输出与基于 Java 的内部处理之间的黏合剂，功能非常强大。

2.4.2 Struts2 OGNL 的使用

Struts2 在 OGNL 之上提供的最大特性就是支持 ValueStack。在 OGNL 上下文中只能有一个根对象，Struts2 的值栈允许存在多个虚拟根对象，如图 2-9 所示。

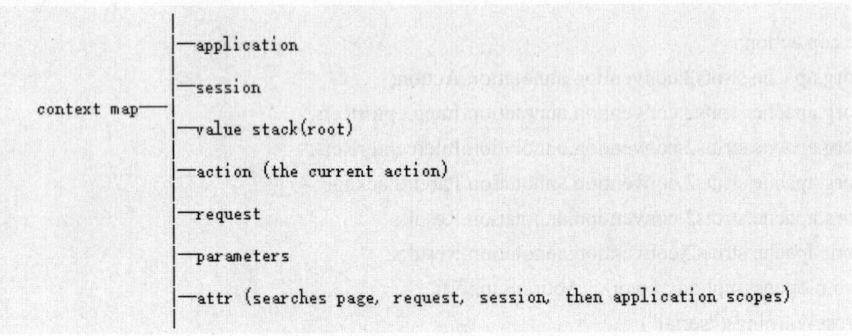

图 2-9 OGNL 根对象

2.4.3 OGNL 访问对象

1. 使用 OGNL 访问 ValueStack

在 OGNL 中，无前缀表示访问当前值栈。例如书写<s:property value="msg"/>，这句话中<s:property>标签的 value 属性的值就是使用的 OGNL，它没有任何前缀，就表示直接访问值栈。

访问到值栈后，会按照从栈顶到栈底的顺序，寻找第一个匹配的对象，那就会找到 Action

中的 msg 属性，然后就可以取到值。

2. OGNL 中的“#”

在 OGNL 中，可以通过符号“#”来访问 ActionContext 中除了值栈之外的各种值，比如下面几个带有“#”的书写习惯：

- #parameters：当前请求中的参数，对应 request.getParameter(name)；
- #request：请求作用域中的属性，对应 request.getAttribute(name)；
- #session：会话作用域中的属性，对应 session.getAttribute(name)；
- #application：应用程序作用域的属性；
- #attr：按照页面 page、请求 request、会话 session 和应用 application 的顺序，返回第一个符合条件的属性。

使用引用需要先加上前缀“#”，并指定范围，然后写出要引用哪个属性，形如：“#paramters.msg”。

下面将通过一个具体的案例讲解 OGNL 访问对象。

（1）在 Eclipse 中新建 Dynamic Web Project 工程 struts5，在 src 的 cap.bean 子包下新建 User.java 类，编辑后的代码如下：

```
package cdavtc.bean;
public class User {
    private int id;
    private String username;
    private String password;
    public User() {}
    public User(String username, String password) {
        this.username = username;
        this.password = password;
    }
    public User(int id, String username, String password) {
        this.id = id;
        this.username = username;
        this.password = password;
    }
    //省略 getters 和 setters
}
```

（2）在 cap.action 子包中新建 OgnlObjAction.java 类，编辑后的代码如下：

```
package cap.action;
import javax.servlet.http.HttpServletRequest;
import org.apache.struts2.ServletActionContext;
import cap.bean.User;
import com.opensymphony.xwork2.ActionContext;
import com.opensymphony.xwork2.ActionSupport;
public class OgnlObjAction extends ActionSupport {
    private User user;
    public String ognlObj() throws Exception {
        // 获得 ActionContext 实例，以便访问 Servlet API
```

```
        ActionContext ctx = ActionContext.getContext();
        // 存入 application
        ctx.getApplication().put("msg", "application 信息");
        // 保存 session
        ctx.getSession().put("msg", "seesion 信息");
        // 保存 request 信息
        HttpServletRequest request = ServletActionContext.getRequest();
        request.setAttribute("msg", "request 信息");
        // 为 User 赋值
        user=new User();
        user.setUsername("cap");
        return SUCCESS;
    }
    public User getUser() {
        return user;
    }
    public void setUser(User user) {
        this.user = user;
    }
}
```

代码解释：在 onglObj 方法中，首先获得 ActionContext 的对象 ctx，通过 ctx 获得 Application 的对象并存储 msg 信息，通过 ctx 获得 Session 对象并存储 msg 信息，通过 ServletActionContext 获得 HttpServletRequest 对象 request，存储 msg 信息到 request 对象中，最后创建了一个 User 对象，并将其属性 username 的值设置为 cap。

（3）在 src 中创建 struts2 的配置文件 struts.xml，编辑后的代码如下：

```
<?xml version="1.0" encoding="UTF-8" ?>
<!DOCTYPE struts PUBLIC
    "-//Apache Software Foundation//DTD Struts Configuration 2.3//EN"
    "http://struts.apache.org/dtds/struts-2.3.dtd">
<struts>
    <package name="default" namespace="/" extends="struts-default">
        <action name="ognlObj" class="cap.action.OgnlObjAction" method="ognlObj">
            <result name="success">/ognlObj.jsp</result>
        </action>
    </package>
</struts>
```

（4）在 WebContent 目录中添加 index.jsp 视图页面，添加下面的代码：

```
<a href="ognlObj?msg=ognl">ONGL 访问对象使用实例</a>
```

（5）在 WebContent 目录中添加 onglObj.jsp 视图页面，编辑后的代码如下：

```
<%@ page language="java" contentType="text/html; charset=UTF-8"
    pageEncoding="UTF-8"%>
<%@ taglib uri="/struts-tags" prefix="s"%>
<!DOCTYPE html PUBLIC "-//W3C//DTD HTML 4.01 Transitional//EN" "http://www.w3.org/TR/html4/loose.
dtd">
```

```
<html>
<head>
<meta http-equiv="Content-Type" content="text/html; charset=UTF-8">
<title>使用 ognl 访问对象</title>
</head>
<body>
        访问 ValueStack 中的对象属性
    <s:property value="user.username"/><!--代码① -->
    <br>
        请求参数中的账号：
      <s:property value="#parameters.msg" /><!--代码② -->
      <br>
      <%
           request.setAttribute("msg", "request 的 msg");
      %>
      请求属性中的账号：
      <s:property value="#request.msg" /><!--代码③ -->
      <br> 会话属性中的账号：
      <s:property value="#session.msg" /><!--代码④ -->
      <br> 应用属性中的账号：
      <s:property value="#application.msg" /><!--代码⑤ -->
      <br> attr 中的信息：
      <s:property value="#attr.msg" /><!--代码⑥ -->
</body>
</html>
```

代码解释：代码①通过 OGNL 直接访问 ValueStack 中的对象属性值。代码②获得 index.jsp?msg=ognl 页面传递过来的参数值。代码③～⑥使用#符号访问 ActionContext 中除了值栈之外的各种值。

（6）运行 struts4：单击 index.jsp 中的超链接，会出现下面的结果，如图 2-10 所示。

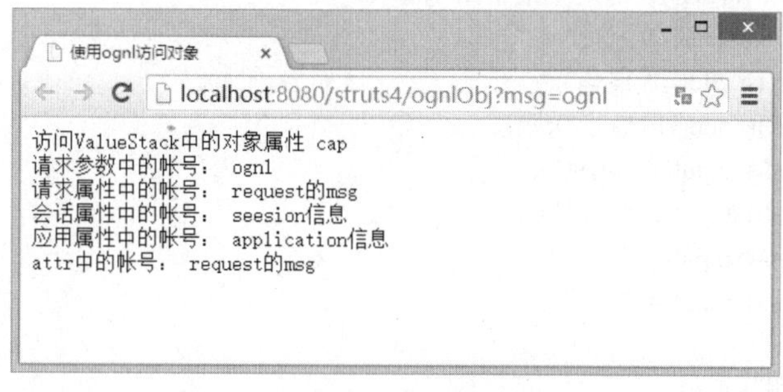

图 2-10　OGNL 访问对象的运行结果

2. OGNL 的集合操作

OGNL 支持各种复杂的表达式。但是最基本的表达式是将对象的引用值用“.”串联起来。从左到右，每一次表达式计算返回的结果成为当前对象，后面部分接着在当前对象上进行计算，一直到全部表达式计算完成，返回最终计算得到的对象。OGNL 针对这条基本原则进行不断的扩充，从而使之支持对象树、数组、容器的访问等操作。

OGNL 能够引用集合的一些特殊的属性，当表达式引用这些属性时，OGNL 会调用相应的方法，这就是伪属性。表 2-3 列出了常用的伪属性。

表 2-3　集合的伪属性

集　合	伪 属　性
Collection(inherited by Map, List & Set)	size，isEmpty
List	iterator
Map	keys，values
Set	iterator
Iterator	next，hasNext
Enumeration	next，hasNext，nextElement，hasMoreElements

下面将通过一个具体的案例来讲解 Struts2 OGNL 对集合的操作。

（1）继续工程 struts4 中 src 的 cap.action 子包中新建类 OgnlSetAction.java，编辑后的代码如下：

```
package cap.action;
import java.util.ArrayList;
import java.util.HashMap;
import java.util.List;
import java.util.Map;
import cap.bean.User;
import com.opensymphony.xwork2.ActionSupport;
public class OgnlSetAction extends ActionSupport{
	private String[] fruits;
	private   List<String> folowersList = new ArrayList<String>();
	private   Map<Integer,String> provincesMap = new HashMap<Integer,String>();
	public OgnlSetAction() {
		fruits=new String[]{"苹果","梨子","水蜜桃"};
		folowersList.add("太阳花");
		folowersList.add("海棠花");
		folowersList.add("风信子");
		provincesMap.put(1, "北京");
		provincesMap.put(2, "上海");
		provincesMap.put(3, "天津");
		provincesMap.put(4, "重庆");
	}
	//省略 getters 和 setters
	public String execute()
	{
		return SUCCESS;
	}
}
```

代码解释：在 OnglSetAction 构造器方法中，主要给 String[] fruits，List<String> folowersList，Map<Integer,String> provincesMap 对象赋初值。

（2）在 Struts2 的配置文件 struts.xml 中添加下面的 action 配置代码：

```
<action name="ognlSet" class="cap.action.OgnlSetAction">
    <result name="success">/ognlSet.jsp</result>
</action>
```

（3）在 WebContent 下的 index.jsp 页面，添加下面的代码：

```
<a href="ognlSet">ONGL 访问集合使用实例</a>
```

（4）在 WebContent 下新建 ognlSet.jsp 页面，编辑后的代码如下：

```
<%@ page language="java" contentType="text/html; charset=UTF-8"
    pageEncoding="UTF-8"%>
<%@taglib uri="/struts-tags" prefix="s" %>
<!DOCTYPE html PUBLIC "-//W3C//DTD HTML 4.01 Transitional//EN" "http://www.w3.org/TR/html4/
loose.dtd">
<html>
<head>
<meta http-equiv="Content-Type" content="text/html; charset=UTF-8">
<title>ognl 示例</title>
</head>
<body>
<!—代码块① -->
<b>数组使用</b>
<br><hr>
<b>fruits :</b> <s:property value="fruits"/> <br>
<b>水果的种类 :</b> <s:property value="fruits.length"/> <br>
<b>第一种水果 :</b> <s:property value="fruits[0]"/> <br>
<br>
<!—代码块② -->
<b>List 使用示例</b>
<br><hr>
<b>folowersList :</b> <s:property value="folowersList"/> <br>
<b>folowersList.size :</b> <s:property value="folowersList.size"/> <br>
<b>folowersList[0] :</b> <s:property value="folowersList[0]"/> <br>
<br>
<!—代码块③ -->
<b>Map 使用示例</b>
<br><hr>
<b>provincesMap[1] :</b> <s:property value="provincesMap[1]"/> <br>
<b>provincesMap.size :</b> <s:property value="provincesMap.size"/> <br>
<br>
</body>
</html>
```

代码解释：代码块①、②、③中主要使用到集合中的各种伪属性，包括数组的 length，List 和 Map 的 size。

（5）在 src 下新建 struts2 的配置 struts.xml，编辑后的代码如下：

```
<!DOCTYPE struts PUBLIC
"-//Apache Software Foundation//DTD Struts Configuration 2.0//EN"
"http://struts.apache.org/dtds/struts-2.0.dtd">
<struts>
    <package name="default" extends="struts-default">
        <action name="ognl" class="cdavtc.action.OgnlAction">
            <result name="success">/ognl.jsp</result>
        </action>
    </package>
</struts>
```

（6）运行结果请读者自行运行工程。

第3章 Struts2视图标签

在早期的Web应用开发过程中，表现层的JSP页面主要使用Java脚本来控制输出。在这种方式下，JSP页面里就会嵌套大量的Java脚本，并且通过Java语言里的if条件语句、for循环语句、while循环语句……来控制输出。这种方式的结果是导致JSP页面里几乎都是Java语言的子集。

当JSP页面里嵌套了大量的Java脚本时，整个页面的可读性会下降，可维护性也随之下降。即使是在前期的开发阶段，页面美工人员因不懂Java语言，故无法参与JSP页面的开发；然而懂Java语言的人员，却不太懂页面的美工设计。因此，大量嵌套Java脚本的JSP技术是不利于团队协作开发的。

从JSP规范1.1版本以后，JSP增加了自定义标签库的规范，自定义标签库是一种非常优秀的组件技术。通过使用自定义标签库，可以在简单的标签中封装复杂的功能，即我们可以在自定义标签中封装复杂的表现逻辑，从而避免在JSP嵌套Java脚本。

Struts2内建的标签库简化了JSP页面输出逻辑的实现，借助于Struts2的标签库，完全可以避免在JSP页面中使用Java脚本，提高表现层组件的可维护性。

Struts2标签API通过使用条件呈现以及集成ValueStack程序的域模型数据，提供了动态创建健壮网页的功能。Struts2自带了很多不同类型的标签，大概分为4种类别：数据标签（data tag）、流程控制标签（control-flow tag）、UI标签（UI tag）和其他标签（miscellaneous tag）。

Struts2标签的JSP版本与其他JSP标签一样。下面是property标签的简单用法。

```
<s:property value="name" />
```

需要注意的是，在使用Struts2标签之前，在页面的开始部分必须定义包含标签库(taglib)的声明。下面的代码片段展示了声明标签库所需要的内容。

```
<%@ page language="java" contentType="text/html; charset=UTF-8"
    pageEncoding="UTF-8"%>
<%@taglib uri="/struts-tags" prefix="s"%>
```

第 3 行 taglib 指令声明了 Struts2 标签库，并且指定“s”作为区分这个标签库的标签前缀。

3.1 数据标签

数据标签能从 ValueStack 上取得数据，或者将变量、对象放在 ValueStack 上。在本节中，我们会讨论 property、push、bean 和 action 标签。本节讲解的标签及其作用如表 3-1 所示。

表 3-1 常用数据标签及其作用

标 签 名	作 用
property	property 标签提供了一种将属性写入呈现的 HTML 页面的快速、方便的方法
bean	bean 标签创建一个对象的实例，把它放到 ValueStack 或者设置为 ActionContext 的顶级引用
push	push 标签允许把属性放到 ValueStack 栈顶
param	用来为其他标签提供参数,一般是嵌套在其他标签的内部

下面将以实际案例来讲解数据标签的使用。

（1）在 Eclipse 中新建 Dynamic Web Project 工程 struts6，复制工程 struts3 中的 User.java 到 src 的 cap.bean 子包下。

（2）在 WebContent 下新建 datatag.jsp 页面，编辑后的代码如下：

```
<%@ page language="java" contentType="text/html; charset=UTF-8"
    pageEncoding="UTF-8"%>
<%@ taglib uri="/struts-tags" prefix="s" %>
<!DOCTYPE html PUBLIC "-//W3C//DTD HTML 4.01 Transitional//EN" "http://www.w3.org/TR/html4/
loose.dtd">
<html>
<head>
<meta http-equiv="Content-Type" content="text/html; charset=UTF-8">
<title>数据标签的使用</title>
</head>
<body>
<s:bean name="cap.bean.User" id="u">
    <s:param name="id" value="29"/>
    <s:param name="username" value="'cap'" />
    <s:param name="password" value="'cap'" />
</s:bean>
<!-- 使用 push 标签将 Stack Context 中的 u 实例放入 ValueStack 栈顶 -->
<s:push value="#u">
    <!-- 输出 ValueStack 栈顶元素的属性 -->
    <s:property value="id"/><br/>
    <s:property value="username" /><br />
    <s:property value="password" /><br />
</s:push>
```

```
</body>
</html>
```

代码解释：在<s:bean>标签中，首先初始化一个User对象u，并设置其属性值。<s:push>标签将对象u实例放入ValueStack的栈顶，然后输出其属性。

3.2　控制标签

Struts2中有一系列的标签可以很容易地控制页面执行的流程，例如分支判断、循环遍历等操作，从而在页面中呈现使用条件的强大功能。常用控制标签及其作用如表3-2所示。

表3-2　控制标签及其作用

标　　签	作　　用
iterator	使用iterator标签可以非常容易地遍历集合对象。主要的属性包含value指定被迭代的集合，id指定集合里面元素的id，可以和var属性替换，status表示迭代元素的索引
if/else/elseif	用于控制逻辑

下面将通过一个案例来介绍控制标签的具体使用。

（1）继续在工程struts6中编写，在WebContent下新建index.jsp页面，在body标签中写入下面的代码：

```
<a href="control.action">control tag</a>
```

（2）在src的cap.action子包下新建ControlAction.java类，编辑后的代码如下：

```
package cap.action;
import java.util.ArrayList;
import java.util.HashMap;
import java.util.List;
import java.util.Map;
import cap.bean.User;
import com.opensymphony.xwork2.ActionSupport;
public class ControlAction extends ActionSupport {
    private Map<String, String> strMap = new HashMap<String, String>();
    private Map<String, User> userMap = new HashMap<String, User>();
    private List<User> usersList=new ArrayList<User>();
    public Map<String, String> getStrMap() {
     return strMap;
}
     public void setStrMap(Map<String, String> strMap) {
         this.strMap = strMap;
     }
     public Map<String, User> getUserMap() {
         return userMap;
     }
     public void setUserMap(Map<String, User> userMap) {
```

```
            this.userMap = userMap;
        }
        public List<User> getUsersList() {
            return usersList;
        }
        public void setUsersList(List<User> usersList) {
            this.usersList = usersList;
        }
        public String control()
    {
                // 值为字符串
                strMap.put("1", "cdavtc1");
                strMap.put("2", "cdavtc2");
                strMap.put("3", "cdavtc3");
                // 值为 bean 对象
                User u1 = new User(1, "starlee1", "starlee1");
                userMap.put("one", u1);
                User u2= new User(2, "starlee2", "starlee2");
                userMap.put("two", u2);
                usersList.add(u1);
                usersList.add(u2);
                return SUCCESS;
      }
    }
```

代码解释：在 control 方法中为 strMap，userMap 和 usersList 对象赋初值。

（3）在 WebContent 下新建 tag.jsp，编辑后的页面代码如下：

```
<%@ page language="java" contentType="text/html; charset=UTF-8"
    pageEncoding="UTF-8"%>
<%@taglib uri="/struts-tags" prefix="s" %>
<!DOCTYPE html PUBLIC "-//W3C//DTD HTML 4.01 Transitional//EN" "http://www.w3.org/TR/html4/loose.
dtd">
<html>
<head>
<meta http-equiv="Content-Type" content="text/html; charset=UTF-8">
<title>控制标签的使用</title>
</head>
<body>
<s:debug/>
    迭代循环 Map 取值
    <table >
     <s:iterator value="strMap" var="entry">
     <tr>
     <td>Key</td><td><s:property value="#entry.key"/></td>
     <td>Value</td><td><s:property value="#entry.value"/></td>
     </tr>
     </s:iterator>
     </table>
```

```
        迭代循环取 Map 对象值
        <table>
          <s:iterator value="userMap" var="entry" status="st">
            <tr>
            <s:if test="#st.Odd">
            <td><s:property value="#entry.value.id"/></td>
              <td><s:property value="#entry.value.username"/> </td>
            <td><s:property value="#entry.value.password"/></td>
            </s:if>
          </tr>
          </s:iterator>
        </table>
    迭代循环取 List 对象值
        <table>
          <s:iterator value="usersList" var="u" status="st">
            <tr>
            <td><s:property value="#u.id"/></td>
              <td><s:property value="#u.username"/> </td>
            <td><s:property value="#u.password"/></td>
          </tr>
          </s:iterator>
        </table>
</body>
</html>
```

代码解释：在第一个 table 标签中使用<s:iterator>标签迭代显示 Map<String, String> strMap 中的值，属性 value 是需要迭代的集合，var 存储每次迭代的对象 id，采用<s:property>标签访问规则是<s:property value="#entry.key"/>和<s:property value="#entry.value"/>。

在第二个 table 中使用<s:iterator>标签迭代显示 Map<String, User> userMap 中的值（注意其中 value 对应的是 User 对象），属性 value 是需要迭代的集合，var 存储每次迭代的对象 id，采用<s:property>标签访问规则是<s:property value="#entry.value.id"/>和<s:property value="#entry. value.username"/>。

<s:if test="#st.Odd">用于判断迭代元素的属性如果为 Odd，则显示其内容，如果为 Even,则不显示。

第三个 table 中使用使用<s:iterator>标签迭代显示 List<User> usersList 中的值，属性 value 是需要迭代的集合，var 存储每次迭代的对象 id，用<s:property>标签访问规则是<s:property value="#u.id"/>等。

（4）修改 src 目录下的 struts.xml，在 package 标签中添加下面的代码：

```
<action name="control" class="cap.action.ControlAction" method="control">
        <result name="success">/tag.jsp</result>
</action>
```

3.3 UI 标签

UI 标签主要用于生成 Web 页面，或者为 Web 页面提供某些功能支持。每一个 UI 组件都是一个功能单元，用户通过 UI 组件与应用程序交互，向应用程序输入数据。任何一个 Struts2 UI 组件的核心都是一个 HTML 表单控件，例如文本框或者下拉列表等。但这些组件不仅仅是一个表示 HTML 输入元素的标签，它们是高层组件，其中的 HTML 元素只是在浏览器中的表现。UI 组件标签与框架的各个部分紧密集成在一起，从数据转移和数据验证到国际化和界面外观。有一些组件甚至联合一个或多个 HTML 表单元素为页面。

具体说来，UI 组件能够做以下事情：

- 生成 HTML 标记;
- 绑定 HTML 表单字段和 Java 端属性;
- 与框架提供的类型转换关联起来;
- 与框架提供的验证关联起来;
- 与框架提供的国际化功能关联起来。

Struts2 中 UI 标签的表单标签大致分为两种：form 标签本身和单个表单标签。

1）form 标签

form（表单）标签是 UI 组件中最重要的一个组件。form 标签为 Struts2 应用程序提供了重要的切入点，其他 UI 组件与服务器交互数据都必须要放在该标签内，它是实现交互页面的重要保证。form 标签的常用属性如表 3-3 所示。

表 3-3　form 标签的属性

属　性	类　型	描　述
action	String	表单提交的目标，可以是 Struts2 动作的名字或者 URL
namespace	String	Struts2 命名空间，在这个命名空间中查找命名动作，或者从这里开始构建 URL，默认为当前命名空间
method	String	与 HTML form 属性相同，默认是 POST
target	String	与 HTML form 属性相同
enctype	String	上传文件时设置为 multipart/form-data
validate	Boolean	与验证框架一起使用，打开客户端 JavaScript 验证

上述的标记很简单，其中 action 属性最重要，这个动作是这个表单提交的目标，其他的所有属性，框架可以提供智能默认值。

2）其余单个表单标签，常用的如表 3-4 所示。

表 3-4　常用表单标签

标　签	作　用
textfield	用于生成文本输入字段
password	用于生成密码输入字段
textarea	用于生成一个基于 HTML 的 textarea 元素的组件

续表

标　签	作　用
checkbox	checkbox 复选框组件使用单个 HTML 的 checkbox 元素创建了一个 Boolean 型的组件
select	这个组件构建在 HTML 下拉列表上，这个组件允许用户从一系列选项中选择一个值
radio	radio 组件提供了与 select 组件很相似的功能，但是 radio 组件不允许选择多个选项
checkboxlist	该组件与 checklist 组件也很相似，它允许选择多个选项
hidden	用于在表单中嵌入隐藏的请求参数而不展示给用户

下面将通过一个综合案例来讲解上述的 UI 组件。

（1）继续在工程 struts6 上操作。修改 index.jsp 页面，在 body 标签中添加下面的代码：

```
<a href='<s:url action="fill"></s:url>'>注册</a>
```

（2）在 src 的 cap.bean 目录下新建 Province.java 类。编辑后的代码如下：

```
package cap.bean;
public class Province {
	private int provinceID;
	private String provinceName;
public Province() {}
	public Province(int provinceID, String provinceName) {
		this.provinceID = provinceID;
		this.provinceName = provinceName;
	}
	//省略 getters 和 setters
}
```

（3）在 WebContent 下新建 register.jsp 页面。编辑后的页面代码如下：

```
<%@ page language="java" contentType="text/html; charset=utf-8"
	pageEncoding="utf-8"%>
<%@taglib uri="/struts-tags" prefix="s"%>
<!DOCTYPE html PUBLIC "-//W3C//DTD HTML 4.01 Transitional//EN" "http://www.w3.org/TR/html4/loose.
dtd">
<html>
<head>
<meta http-equiv="Content-Type" content="text/html; charset=utf-8">
<title>Register Page</title>
</head>
<body>
<s:form action="register" method="post">
	<s:textfield name="userName" label="用户名" />
	<s:password name="password" label="密码" />
	<s:radio name="gender" label="性别" list="{'Male','Female'}" />
	<s:select name="province" list="provinceList" listKey="provinceID"
		listValue="provinceName" label="选择省份" />
	<s:textarea name="about" label="自己介绍" />
	<s:checkboxlist list="skiList" name="skills" label="编程技能" />
```

```
        <s:checkbox name="confirm" label="同意注册?" />
        <s:submit value="注册"/>
</s:form>
</body>
</html>
```

（4）在 src 的 cap.action 子包下新建 RegisterAction.java 类。编辑后的代码如下：

```
package cap.action;
import java.util.ArrayList;
import com.opensymphony.xwork2.ActionSupport;
import cap.bean.Province;
public class RegisterAction extends ActionSupport{
        private String userName;
        private String password;
        private String gender;
        private String about;
        private String province;
        private ArrayList<Province> provinceList;
        private String[] skills;
        private ArrayList<String> skiList;
        private Boolean confirm;
        public String fill() {
                provinceList = new ArrayList<Province>();
                provinceList.add(new Province(1, "北京"));
                provinceList.add(new Province(2, "上海"));
                provinceList.add(new Province(3, "天津"));
                provinceList.add(new Province(3, "重庆"));
                skiList = new ArrayList<String>();
                skiList.add("C");
                skiList.add("Java");
                skiList.add(".Net");
                skiList.add("C++");
                skills = new String[]{"Java",".Net"};
                confirm = true;
                return "fill";
        }
        public String register() {
                return SUCCESS;
        }
        //省略 getters 和 setters
}
```

代码解释：fill 方法用于初始化各种变量，包括 provinceList，skiList 等对象。

（5）在 WebContent 下新建 result.jsp 页面。编辑后的代码如下：

```
<%@ page language="java" contentType="text/html; charset=utf-8"
pageEncoding="utf-8"%>
<%@taglib uri="/struts-tags"    prefix="s"%>
```

```
<!DOCTYPE html PUBLIC "-//W3C//DTD HTML 4.01 Transitional//EN" "http://www.w3.org/TR/html4/loose.dtd">
<html>
<head>
<meta http-equiv="Content-Type" content="text/html; charset=utf-8">
<title>Details Page</title>
</head>
<body>
<s:debug/>
用户名: <s:property value="userName" /><br>
性别: <s:property value="gender" /><br>
省份: <s:property value="province" /><br>
关于自己: <s:property value="about" /><br>
技能: <s:property value="skills" /><br>
确认: <s:property value="confirm" />
</body>
</html>
```

（6）编辑 src 的 struts.xml 文件，添加下面的代码内容：

```
<action name="register" method="register" class="cap.action.RegisterAction">
        <result name="fill">/register.jsp</result>
        <result name="input">/register.jsp</result>
        <result name="success">/result.jsp</result>
</action>
<action name="fill" method="fill" class="cap.action.RegisterAction">
        <result name="fill">/register.jsp</result>
        <result name="input">/register.jsp</result>
</action>
```

（7）继续运行工程 struts6。在出现的如图 3-1 所示的页面中录入相关信息，单击“注册”按钮，可见如图 3-2 所示的注册结果。

图 3-1 注册页面

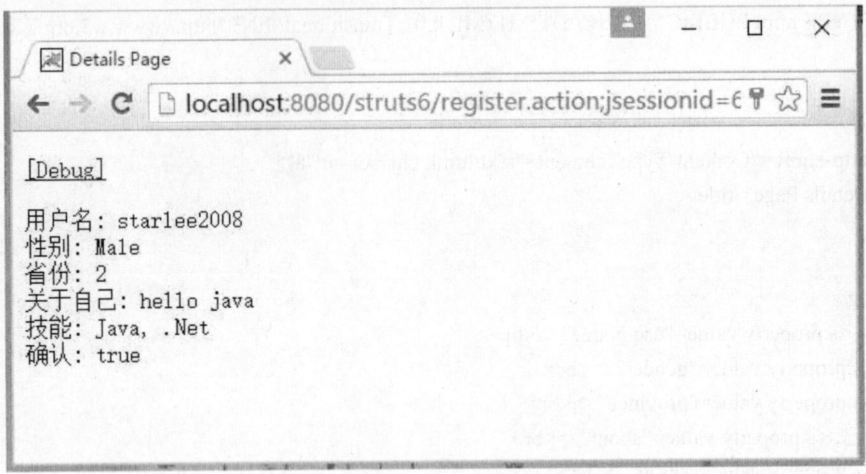

图 3-2　注册的结果

3.4　其他标签

除了上面讲到的数据库标签，控制标签，UI 标签，还有一些标签不方便归类，统一把它分为其他标签，在本节中主要讲到的标签如表 3-5 所示。

表 3-5　其他常用标签

标　　签	作　　用
<s:url>	用于创建一个 URL 链接，可以通过<s:param>标签传递参数
<s:action>	可以直接调用 Action 类中的函数，前提是该 Action 类在 struts.xml 中有定义，name 属性用于指向映射文件中配置的 Action 函数，executeResult 属性表示是否执行结果集的跳转
<s:a>	用于创建一个<a>标签，errorText 用于显示错误信息，priInvokeJS 表示该链接执行时的 JavaScript 提示函数，href 表示跳转地址

下面将通过一个实例来讲解上述的其他常用标签。

（1）继续在工程 struts6 上操作。新建 other_surl.jsp 页面，编辑后的代码如下：

```
<%@ page language="java" contentType="text/html; charset=UTF-8"
    pageEncoding="UTF-8"%>
<%@ taglib uri="/struts-tags" prefix="s" %>
<!—省略部分 HTML 标签  -->
<body>
<!--会生成 index.jsp?username=starlee2008 的链接   -->
<s:url id="url" value="/index.jsp" var="url">
<s:param name="username">starlee2008</s:param>
</s:url>
<s:a href="%{url}">s:url 和 s:a 的用法</s:a>
</body>
</html>
```

（2）在 src 的 cap.action 中新建 LoginAction.java 类，编辑后的代码如下：

```
package cap.action;
import com.opensymphony.xwork2.ActionSupport;
public class LoginAction extends ActionSupport{
        public String login(){
                System.out.println("login");
                return SUCCESS;
                  }
}
```

（3）在 src 的 Struts2 的配置文件 struts.xml 中添加 action 的配置代码：

```
<action name="login"   class="cap.action.LoginAction" method="login">
                <result name="success">/index.jsp</result>
</action>
```

（4）继续在 WebContent 中新建 other_saction.jsp 页面，添加下面的代码：

```
<%@ page language="java" contentType="text/html; charset=UTF-8"
    pageEncoding="UTF-8"%>
<%@ taglib uri="/struts-tags" prefix="s" %>
<!—省略部分 HTML 标签   -->
<body>
<!--表示执行 LoginAction 的 login 方法   -->
<s:action name="login" namespace="/" executeResult="true" />
</body>
</html>
```

（5）本节中列举的标签用法比较简单，页面的运行结果请读者自行验证。

第4章 Struts2 国际化和数据校验

4.1 国际化

要想让一个应用在全球范围内都能使用，就需要充分考虑不同地域和语言环境的因素，最基本应该实现的就是用户界面上的信息可以用本地化语言来显示。当然，一个优秀的全球化软件产品，对国际化和本地化的要求远远不止于此，甚至还包括用户提交数据的国际化和本地化。

Java 语言内核基于 Unicode 2.1 标准，提供了对不同国家和不同语言文字的内部支持。这也是 Java 对于国际化的支持远比 C/C++优越的原因所在。

国际化是指应用程序运行时，可根据客户端请求来自的国家/地区、语言的不同而显示不同的界面。例如，如果请求来自于中文操作系统的客户端，则应用程序中的各种提示信息、错误和帮助等都会使用中文文字；如果客户端使用英文操作系统，则应用程序能自动识别，并做出英文的响应。

引入国际化的目的是为了提供自适应、更友好的用户界面，而并未改变程序的逻辑功能。国际化的英文单词是 Internationalization，简称 i18n（I18N），其中 i 是这个单词的第一个字母，18 表示这个单词的长度，而 n 代表这个单词的最后一个字母。

一个国际化支持很好的应用，会随着在不同区域使用时，呈现出不同区域相应的语言提示。因此，有时候这个过程也被称为 Localization，即本地化。类似于国际化可以称为 i18n，本地化也可以称为 l10n（L10N）。

Struts2 框架为 i18n 付出了很多努力。框架仍然使用我们刚才看过的 Java 类，但应用实现更容易。首先，不需要实例化 ResourceBundle，Struts2 自动创建 ResourceBundle，并确定处理需要哪个包的所有繁琐内容。另外，框架也处理确定正确地域。框架通过检查来源于浏览器的 HTTP 头信息自动确定当前的 Locale。如果愿意，你也可以覆盖这个行为，并使用其他方式确定地域，例如让用户通过基于用户界面的选择来手动选取一个地域口。

4.1.1 加载国际化资源

Struts2 加载国际化资源最简单、最常用的方式就是加载全局的国际化资源文件，该方式主要通过配置常量来实现。只需要在 struts.xml 中配置如下语句即可。

```
<constant name="struts.custom.i18n.resources" value="message"/>
```

通过这种方式加载国际化资源，在 JSP 和 Action 中都可以访问。

4.1.2 访问国际化消息

Struts2 既可以在 JSP 页面中通过标签来输出国际化消息，也可以在 Action 类中输出国际化消息。不管采用哪种方式，Struts2 都提供了非常简单的支持。Struts2 访问国际化消息主要有下面三种方式。

第一，为了在 JSP 页面输出国际化消息，可以使用 Struts2 的标签<s:text name=" "/>，该标签的 name 属性就指定了国际化资源文件中的 key。

第二，为了在 Action 类中访问国际化消息，可以使用 ActionSupport 类的 getText 方法，该方法可以接受一个 name 参数，该参数对应国际化资源文件中的 key。

第三，为了在表单元素的 label 里输出国际化信息，可以为该表单标签的 key 属性赋值，该属性对应国际化资源文件中的 key。

4.1.3 国际化案例

（1）打开 Eclipse IDE，安装插件 Properties Editor。推荐插件下载地址：http://propedit.sourceforge.jp/eclipse/updates，详细的安装查看附录 A Eclipse 插件的安装。

（2）在 Eclipse 中新建 Dynamic Web Project 工程 struts7，在 src 下新建国际化资源文件 message.properties，使用 Properties Editor 打开编辑后的内容如下。

```
register.page=用户注册
register.title=新用户注册
username=用户名
password=密码
repassword=确认密码
submit=立即注册
name.null=用户名不能为空
name.length=用户名长度必须在 6 和 16 之间
password.null=密码不能为空
password.length=密码长度必须大于等于 8
repassword.null=确认密码不能为空
repassword.same=密码和确认密码必须相同
```

（3）在 src 下新建美国英语的国际化语言 message_en_US.properties，编辑后的代码如下：

```
register.page=User Register
register.title=New User Register
username=Name
password=Password
```

```
repassword=RePassword
submit=Register Now
name.null=Name cannot be null
name.length=Name should be between 6 and 16
password.null=Password cannot be null
password.length=Mininum password length is 8
repassword.null=Repassword cannot be null
repassword.same=Repassword should be same with password
```

（4）在 WebContent 下新建 register.jsp 页面，编辑后的代码如下：

```
<%@ page language="java" contentType="text/html; charset=UTF-8"
    pageEncoding="UTF-8"%>
<%@taglib prefix="s" uri="/struts-tags"%>
<!DOCTYPE html PUBLIC "-//W3C//DTD HTML 4.01 Transitional//EN" "http://www.w3.org/TR/html4/loose.
dtd">
<html>
<head>
<meta http-equiv="Content-Type" content="text/html; charset=UTF-8">
<title><s:text name="register.page" /></title>
</head>
<body>
<h2>
        <s:text name="register.title" />
        </h2>
        <s:form action="register" method="post">
            <s:textfield name="user.name" key="username"    />
            <s:password name="user.password" key="password" />
            <s:password name="repassword" key="repassword" />
            <s:submit key="submit" />
        </s:form>
</body>
</html>
```

（5）在 src 下新建 struts 的配置文件 struts.xml，编辑后的代码如下：

```
<?xml version="1.0" encoding="UTF-8" ?>
<!DOCTYPE struts PUBLIC
    "-//Apache Software Foundation//DTD Struts Configuration 2.0//EN"
    "http://struts.apache.org/dtds/struts-2.0.dtd">
<struts>
    <constant name="struts.custom.i18n.resources" value="message"/>
    <constant name="struts.i18n.encoding" value="utf-8"></constant>
    <package name="default" namespace="/" extends="struts-default">
  </package>
</struts>
```

代码解释：第 1 个<constant>标签用于设置国际化消息的名字，第 2 个<constant>标签将国际化的编码设置为 UTF-8。

（6）运行工程：打开 Internet Explorer 中的 Internet 选项，设置语言，在弹出的对话框中选择设置语言首选项，将英语设置为第一语言，可以在 IE 中看见如图 4-1 所示的运行结果：

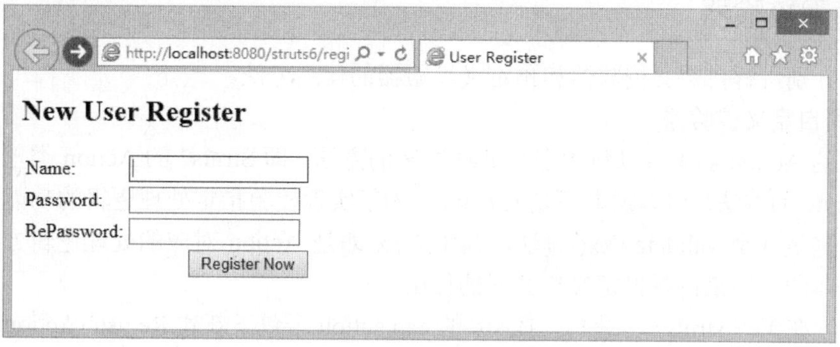

图 4-1　国际化运行结果

4.2　Struts2 校验框架

输入校验是所有 Web 应用应当积极处理的问题，因为 Web 应用的开放性，网络上所有的浏览者都可以自由使用该应用，因此该应用通过输入页面收集的数据是非常复杂的，不仅有正常用户的正确输入，还可能包含例如正常用户的误输入或恶意用户的恶意输入等数据。一个健壮的应用系统应该将这些非法输入阻止在应用系统之外，防止这些非法输入进入系统，这样才可以保证系统不受影响。

输入校验分为客户端校验和服务器校验，客户端校验的目的是要过滤正常用户的误操作，主要通过 JavaScript 代码完成，服务器端校验是整个应用阻止非法数据的最后防线，主要通过在应用中编程实现。

Struts2 框架提供了非常强大的输入校验体系，通过 Struts2 内建的输入校验器，Struts2 应用无需书写任何输入校验代码，即可完成绝大部分输入校验，并可以同时完成客户端校验和服务器端校验。如果应用的输入校验规则特别完整全面，Struts2 也允许通过重写 validate 方法来完成自定义校验。除此之外，Struts2 的开放性还允许开发者提供自定义的校验器。

4.2.1　验证框架

只要 Struts2 Action 类继承 ActionSupport 类，这个类就实现了两个在验证中扮演重要角色的接口。这两个接口分别是 com.opensymphony.xwork2.Validateable 和 com.opensymphony.xwork2.ValidationAware。Validateable 接口提供了 validate()方法，在这个方法中放入了数据验证的代码，ValidationAware 公开了当验证发现无效数据时用来存储生成的错误消息的方法。这两个接口与 workflow 拦截器一起协作。当 workflow 拦截器触发时，它首先检查动作是否实现 Validateable 接口。如果实现了这个接口，workflow 拦截器调用 validate()方法。如果验证代码发现一些数据不合法，会创建一个错误消息并且添加到 ValidationAware 的某个保存错误消息的方法中。当 validate()方法返回时，workflow 拦截器会调用 ValidationAware 的 hasErrors()方法来查看是否有任何验证错误，如果存在错误消息，workflow 拦截器会干涉并停止动作的后续执行并返回输入结果，这个结果将带回用户提交请求的表单。

4.2.2 使用校验器

下面将分别讲解自定义校验器和预定义校验器的使用方法。

1. 使用自定义校验器

Struts2 的 Action 类里可以包含多个处理逻辑的方法。即 Struts2 的 Action 类里定义了几个类似于 execute 的方法，但方法名不是 execute。为了实现校验指定处理逻辑的功能，Struts2 的 Action 允许提供一个 validateXxx()方法，其中 Xxx 即是 Action 对应的处理逻辑方法。下面将通过一个具体的案例来讲解自定义拦截器的使用。

（1）继续在工程 struts7 中编写，在 src 的 cap.action 子包下新建 RegisterAction.java 类，编辑后的代码如下：

```
package cap.action;
import cap.bean.User;
import com.opensymphony.xwork2.ActionSupport;
public class RegisterAction extends ActionSupport {
        private static final long serialVersionUID = 1L;
        private User user;
        public void validateRegister(){
                if(user.getUsername()==null || user.getUsername().equals("")){
                        //this.addFieldError("user.name","用户名不能为空");
                        this.addFieldError("user.name",getText("name.null"));
                }
        }
        public String register()
        {
                return SUCCESS;
        }
        public User getUser() {
                return user;
        }
        public void setUser(User user) {
                this.user = user;
        }
}
```

代码解释：validateResister()方法对应于 Action 中的 regiscer()方法，其用于验证用户名，如果用户名为 null 或者为空字符串，将会调用 addFieldError 方法添加错误信息。

（2）在 src 的 cap.bean 子包下新建 User.java 类，代码如下：

```
package cdavtc.vo;
public class User {
        private int id;
        private String username;
        private String password;
        private String repassword;
```

```
    //省略 getters 和 setters
    }
```

（3）在 src 下修改 struts.xml，在 package 标签中添加下面的内容：

```
<action name="register" class="cap.action.RegisterAction" method="register">
        <result name="success">/index.jsp</result>
        <result name="input">/register.jsp</result>
</action>
```

（4）运行工程后，当用户在没输入任何数据的情况下单击“立即注册”按钮，会出现如图 4-2 所示的提示。

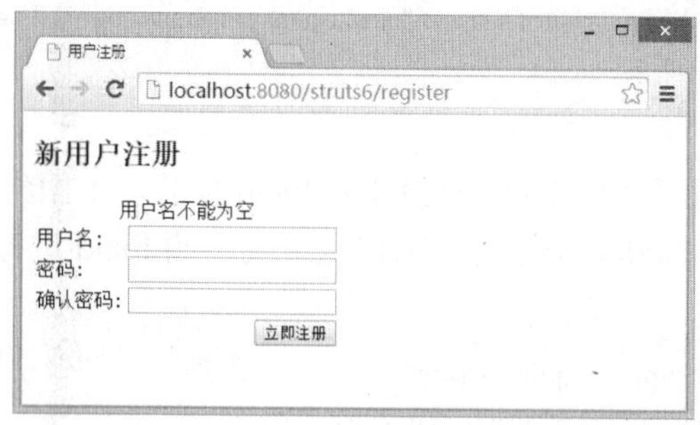

图 4-2　自定义校验器的运行结果

2. 使用预定义校验器

采用 Struts2 的校验框架时，只需要为该 Action 指定一个校验文件即可。校验文件是一个 XML 配置文件，该文件指定了 Action 的属性必须满足怎样的规则。该 XML 文件取名应该遵守如下规则。

```
Action 类名-validation.xml
```

Struts2 已经实现很多常用的校验了，如下面代码所示：

```
<?xml version="1.0" encoding="UTF-8"?>
<!DOCTYPE validators PUBLIC
        "-//Apache Struts//XWork Validator Definition 1.0//EN"
        "http://struts.apache.org/dtds/xwork-validator-definition-1.0.dtd">
<validators>
    <validator name="required" class="com.opensymphony.xwork2.validator.validators.RequiredFieldValidator"/>
    <validator name="requiredstring" class="com.opensymphony.xwork2.validator.validators.
RequiredStringValidator"/>
    <validator name="int" class="com.opensymphony.xwork2.validator.validators.IntRangeFieldValidator"/>
    <validator name="long" class="com.opensymphony.xwork2.validator.validators.LongRangeFieldValidator"/>
    <validator name="short" class="com.opensymphony.xwork2.validator.validators.
ShortRangeFieldValidator"/>
    <validator name="double" class="com.opensymphony.xwork2.validator.validators.
DoubleRangeFieldValidator"/>
    <validator name="date" class="com.opensymphony.xwork2.validator.validators.
DateRangeFieldValidator"/>
    <validator name="expression" class="com.opensymphony.xwork2.validator.validators.
```

```
ExpressionValidator"/>
        <validator name="fieldexpression" class="com.opensymphony.xwork2.validator.validators.
FieldExpressionValidator"/>
        <validator name="email" class="com.opensymphony.xwork2.validator.validators.EmailValidator"/>
        <validator name="url" class="com.opensymphony.xwork2.validator.validators.URLValidator"/>
        <validator name="visitor" class="com.opensymphony.xwork2.validator.validators.VisitorFieldValidator"/>
        <validator name="conversion" class="com.opensymphony.xwork2.validator.validators.
ConversionErrorFieldValidator"/>
        <validator name="stringlength" class="com.opensymphony.xwork2.validator.validators.
StringLengthFieldValidator"/>
        <validator name="regex" class="com.opensymphony.xwork2.validator.validators.RegexFieldValidator"/>
        <validator name="conditionalvisitor" class="com.opensymphony.xwork2.validator.validators.
ConditionalVisitorFieldValidator"/>
    </validators>
```

校验规则主要包括下面两种，Field 检验和非 Field 检验，下面将讲述这两种验证的规则。

（1）Field 校验：针对 Action 类中每个非自定义类型的 Field 进行校验的规则。

```
<field name="user.name">
        <field-validator type="requiredstring">
            <param name="trim">true</param>
            <message key="name.null"/>
        </field-validator>
<!—可以添加其他的检验规则 -->
            </field>
```

（2）非 Field 校验：针对 Action 类的某些 Field 使用 OGNL 表达式进行组合校验。

```
<field name="repassword">
        <!—可以添加其他的检验规则 -->
        <field-validator type="fieldexpression">
            <param name="expression">user.password==repassword</param>
            <message key="repassword.same"/>
        </field-validator>
    </field>
```

现在，我们通过一个具体实例讲解使用 XML 文件实现数据校验。

（1）在工程的 cap.action 中创建 RegisterAction-validation.xml，同时将 RegisterAction 中的 public void validateRegister()方法注释掉，RegisterAction-validation.xml 编辑后的代码如下：

```
<!DOCTYPE validators PUBLIC "-//Apache Struts//XWork Validator 1.0.3//EN"
        "http://struts.apache.org/dtds/xwork-validator-1.0.3.dtd">
<validators>
    <field name="user.name">
        <field-validator type="requiredstring">
            <param name="trim">true</param>
            <message key="name.null"/>
        </field-validator>
        <field-validator type="stringlength">
            <param name="maxLength">10</param>
```

```
                <param name="minLength">6</param>
                <message key="name.length"/>
            </field-validator>
        </field>
        <field name="user.password">
            <field-validator type="requiredstring">
                <message key="password.null"/>
            </field-validator>
            <field-validator type="stringlength">
                <param name="minLength">6</param>
                <message key="password.length"/>
            </field-validator>
        </field>
        <field name="repassword">
            <field-validator type="requiredstring">
                <message key="repassword.null"/>
            </field-validator>
            <field-validator type="fieldexpression">
                <param name="expression">user.password==repassword</param>
                <message key="repassword.same"/>
            </field-validator>
        </field>
    </validators>
```

（2）再次发布运行应用程序，会出现和图 4-2 相似的错误页面。

第 5 章 Struts2 应用

5.1 Struts2 文件上传

Web 应用经常会涉及文件上传，在大部分时候，用户的请求参数是在表单域输入的字符串。如果将表单元素设置 enctype="multipart/form-data" 属性，则提交表单时不再以字符串方式提交请求参数，而是以二进制编码的方式提交请求，此时直接通过 HttpServletRequest 的 getParameter 方法无法正常获取请求参数的值，我们可以通过二进制流来获取请求内容，通过这种方式，就可以取得上传文件的内容，从而实现文件的上传。

上面介绍的文件上传是全手工的文件上传机制，这种方式编程麻烦，需要手工控制二进制流，比较复杂且容易出错。

本节将讨论在 Struts2 中使用 commons-fileupload 实现文件上传的实现问题。这里需要导入实现文件下载上传的两个 jar 文件，一个是 commons-fileupload-x.x.x.jar，另一个是 commons-io-x.x.x.jar(x 代表版本号)。Commons FileUpload 通过将 HTTP 的数据保存到临时文件夹，然后 Struts2 使用 fileUpload 拦截器将文件绑定到 Action 的实例中，从而就能够以对待本地文件的方式操作浏览器上传的文件。

下面将通过具体的案例讲解单文件上传和多文件上传两种方式。

5.1.1 单文件上传

（1）在 Eclipse IDE 中新建 Dynamic Web Project 工程 struts8，工程的结构图如图 5-1 所示。

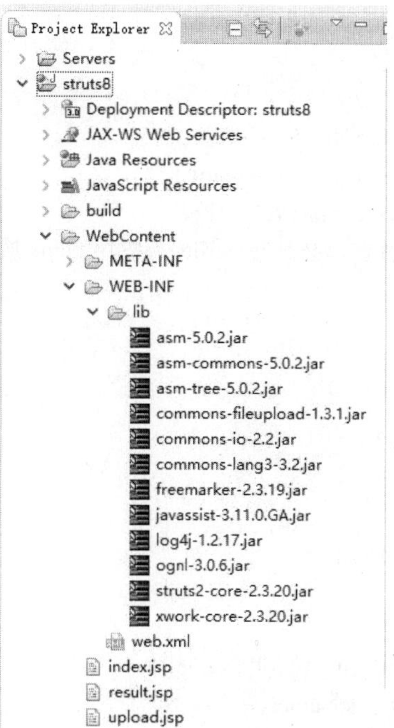

图 5-1 工程 struts8 的结构图

（2）在 WebContent 下新建上传页面，编辑后的代码如下：

```
<%@ page language="java" contentType="text/html; charset=UTF-8"
    pageEncoding="UTF-8"%>
<%@ taglib uri="/struts-tags" prefix="s"%>
<!DOCTYPE html PUBLIC "-//W3C//DTD HTML 4.01 Transitional//EN" "http://www.w3.org/TR/html4/loose.dtd">
<html>
<head>
<meta http-equiv="Content-Type" content="text/html; charset=UTF-8">
<title>文件上传</title>
</head>
<body>
<s:form action="upload" enctype="multipart/form-data">
<s:file name="file"></s:file>
<s:submit value="submit"></s:submit>
</s:form>
</body>
</html>
```

代码解释：在 upload.jsp 中，将表单的提交方式设为 POST，必须将 enctype 设为 multipart/form-data。接下来，<s:file/>标志将文件上传控件绑定到 Action 的 file 属性。

（3）在 src 的 cap.action 子包新建 FileUploadAction.java 类，编辑后的代码如下：

```
package cap.action;
import java.io.File;
import java.io.FileInputStream;
```

```
import java.io.FileOutputStream;
import java.io.InputStream;
import java.io.OutputStream;
import com.opensymphony.xwork2.ActionSupport;
public class FileUploadAction extends ActionSupport {
        private static final long serialVersionUID = 1L;
        private File file;//上传的文件名，必须和<s:file>标签中 name 属性值一致
        private String fileFileName;
        public File getFile() {
                return file;
        }
        public void setFile(File file) {
                this.file = file;
        }
        public String getFileFileName() {
                return fileFileName;
        }
        public void setFileFileName(String fileFileName) {
                this.fileFileName = fileFileName;
        }
        public String execute() throws Exception
        {
                InputStream is=new FileInputStream(file);
                OutputStream os=new FileOutputStream("d:\\"+fileFileName);
                byte[] buffer=new byte[1024];
                int length=0;
                while((length=is.read(buffer))!=-1){
                        os.write(buffer, 0, length);
                }
                os.close();
                is.close();
                return SUCCESS;
        }
}
```

代码解释：在 execute 方法中首先使用 FileInputStream 读取需要上传的文件，然后通过 FileOutStream 将字节流写入到需要保存的文件中，使用完毕后记得关闭输入/输出流。

（4）在 WebContent 下创建文件上传成功的页面 result.jsp。编辑后的代码如下：

```
<%@ page language="java" contentType="text/html; charset=UTF-8"
    pageEncoding="UTF-8"%>
<%@ taglib uri="/struts-tags" prefix="s"%>
<!DOCTYPE html PUBLIC "-//W3C//DTD HTML 4.01 Transitional//EN" "http://www.w3.org/TR/html4/loose.
dtd">
<html>
<head>
<meta http-equiv="Content-Type" content="text/html; charset=UTF-8">
<title>文件上传列表</title>
```

```
</head>
<body>
<s:property value="fileFileName"/>
</body>
</html>
```

（5）修改 src 目录下的 struts 配置文件 struts.xml，在 package 标签中添加下面的代码：

```
<action name="upload" class="cap.action.FileUploadAction">
            <result name="success">/result.jsp</result>
</action>
```

（6）发布运行应用程序：在浏览器地址栏中键入：http://localhost:8080/struts8/upload.jsp，出现如图 5-2 所示的页面，选择文件，单击“上传”按钮提交，出现如图 5-3 所示页面。

图 5-2　upload.jsp 运行页面

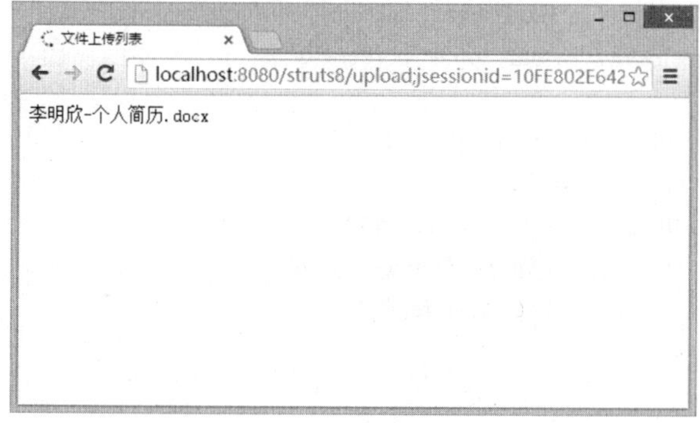

图 5-3　文件上传成功的页面

5.1.2　多文件上传

与单文件上传相似，Struts2 实现多文件上传也很简单。可以将多个<s:file />绑定 Action 的数组或列表。本步骤将开始实现多文件上传，采用 js 来实现动态决定文件的个数，

（1）在 WebContent 下新建 uploads.jsp，编辑后的代码如下：

```
<%@ page language="java" import="java.util.*" pageEncoding="gbk"%>
<%@ taglib prefix="s" uri="/struts-tags"%>
```

```
<!DOCTYPE HTML PUBLIC "-//W3C//DTD HTML 4.01 Transitional//EN">
<html>
<head>
<title>Uploads</title>
<script type="text/javascript">
function addMore(){
        var td = document.getElementById("more");
        var br = document.createElement("br");
        var input = document.createElement("input");
        var button = document.createElement("input");
        input.type = "file";
        input.name = "file";
        button.type = "button";
        button.value = "-";
        button.onclick = function() {
        td.removeChild(br);
        td.removeChild(input);
        td.removeChild(button);
}
                td.appendChild(br);
                td.appendChild(input);
                td.appendChild(button);
        }
</script>
</head>
<body>
        <s:form action="uploads" method="post" enctype="multipart/form-data">
                <table align="center" width="60%" border="1">
                        <tr>
                                <td>选择上传的文件:</td>
                                <td id="more">
                                <input type="file" name="file">
                                <input type="button" value="添加"
                                        onclick="addMore()" />
                                </td>
                        </tr>
                        <tr>
                                <td></td>
                                <td><s:submit value="上传" align="center"></s:submit></td>
                        </tr>
                </table>
        </s:form>
</body>
</html>
```

代码解释：在页面中使用 JavaScript 脚本实现文件添加选项，是通过 addMore 方法实现的，具体的使用参考 JavaScript 方面的知识。

（2）在 src 的 cap.action 子包下新建 FilesUploadAction.java 类，编辑后的代码如下：

```
package cap.action;
import java.io.File;
import java.io.FileInputStream;
import java.io.FileOutputStream;
import java.io.InputStream;
import java.io.OutputStream;
import java.util.List;
import com.opensymphony.xwork2.ActionSupport;
public class FilesUploadAction extends ActionSupport{
    private static final long serialVersionUID = 1L;
    private List<File> file;
    private List<String> fileFileName;
    private List<String> fileContentType;
    public List<File> getFile() {
        return file;
    }
    public void setFile(List<File> file) {
        this.file = file;
    }
    public List<String> getFileContentType() {
        return fileContentType;
    }
    public void setFileContentType(List<String> fileContentType) {
        this.fileContentType = fileContentType;
    }
    public List<String> getFileFileName() {
        return fileFileName;
    }
    public void setFileFileName(List<String> fileFileName) {
        this.fileFileName = fileFileName;
    }
    /**
     * 动态上传文件
     */
    public String execute() throws Exception {

        InputStream is = null;
        OutputStream ops = null;
        for (int i = 0; i < file.size(); i++) {
            try {
                is = new FileInputStream(file.get(i));
                String root = "d:";
                File destFile = new File(root + "\\", this.getFileFileName()
                        .get(i));
                ops = new FileOutputStream(destFile);
                byte[] b = new byte[400];
                int length = 0;
```

```
                    while ((length = is.read(b)) > 0) {
                        ops.write(b, 0, length);
                    }
                } catch (Exception ex) {
                    ex.printStackTrace();
                } finally {
                    is.close();
                    ops.close();
                }
            }
            return SUCCESS;
        }
}
```

（3）在 struts.xml 配置文件中添加下面的 action。

```
<action name="uploads" class="cap.action.FilesUploadAction">
<result name="success">/result.jsp</result>
</action>
```

（4）运行的效果和单文件上传相似，请读者自行运行验证。

5.2 Struts2+JDBC 实现增删改查

“增、删、改、查”是普通应用程序的缩影。如果掌握了增、删、改、查的具体实现，那么意味可以使用该框架创建普通应用程序了，所以大家使用新框架开发应用程序时，首先会研究如何编写 CRUD。本节中使用到的 JDBC 知识可以参看《基于 BootStrap3 的 JSP 项目实例教程》，其余的知识点都在前面已经讲述，所以不再给出代码解释。

下面将通过具体的案例讲解 Struts2 结合 JDBC 实现“增、删、改、查”操作。

（1）在 Eclipse 中新建 Dynamic Web Project 工程 struts9，工程的结构图如图 5-4 所示。

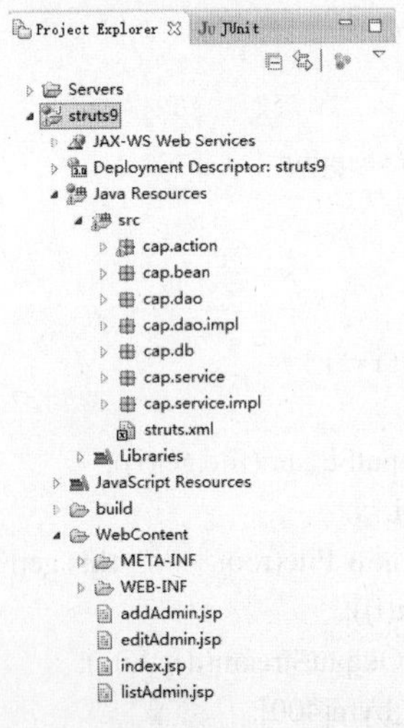

图 5-4　struts9 的工程结构图

（2）在 src 的 cap.db 中新建 DBCon.java 类，编辑后的代码如下：

```
package cap.db;
import java.sql.*;
public class DBCon {
    private static String driver = "com.mysql.jdbc.Driver";
    private String url = "jdbc:mysql://localhost:3306/cap";
    private String user = "root";
    private String password = "admin";// 此处根据自己的实际情况修改为自己设定的密码
    private Connection conn = null;
    public Connection getConnection() {
        try {
            Class.forName(driver);
            conn = DriverManager.getConnection(url, user, password);
        } catch (ClassNotFoundException e) {
            e.printStackTrace();
        } catch (SQLException e) {
            e.printStackTrace();
        }
        return conn;
    }
    public ResultSet doQueryRS(String sql, Object[] params){
        ResultSet rs = null;
        conn = this.getConnection();
        PreparedStatement pstmt;
        try {
            pstmt = conn.prepareStatement(sql);
            for (int i = 0; i < params.length; i++) {
                pstmt.setObject(i + 1, params[i]);
            }
            rs = pstmt.executeQuery();
        } catch (SQLException e) {
            e.printStackTrace();
        }
        return rs;
    }
    public int doUpdate(String sql, Object[] params) {
        int res = 0;
        conn = this.getConnection();
        PreparedStatement pstmt;
        try {
            pstmt = conn.prepareStatement(sql);
            for (int i = 0; i < params.length; i++) {
                pstmt.setObject(i + 1, params[i]);
            }
            res = pstmt.executeUpdate();
        } catch (SQLException e) {
            e.printStackTrace();
```

```
            }
            return res;
        }
        public void close() {
            if (conn != null)
                try {
                    conn.close();
                } catch (SQLException e) {
                    e.printStackTrace();
                }
        }
}
```

（3）在 src 的 cap.bean 子包中新建 Admin.java ，编辑后的代码如下：

```
package cap.bean;
public class Admin {
        private int id;
        private String username;
        private String password;
        //省略 getters 和 setters
}
```

（4）在 src 的 cap.dao 子包中新建 AdminDAO 接口，编辑后的代码如下：

```
package cap.dao;
import java.util.List;
import cap.bean.Admin;
public interface AdminDAO {
        public List<Admin> findAdmins();
        public Admin FindByID(Integer id);
        public int addAdmin(Admin admin);
        public int delAdmin(Integer id);
        public int updateAdmin(Admin admin);
}
```

（5）在 src 的 cap.dao.impl 子包中新建 AdminDAO 接口的实现类 AdminDAOImpl.java，编辑后的代码如下：

```
package cap.dao.impl;
import java.sql.ResultSet;
import java.sql.SQLException;
import java.util.ArrayList;
import java.util.List;
import cap.bean.Admin;
import cap.dao.AdminDAO;
import cap.db.DBCon;
public class AdminDAOImpl implements AdminDAO {
        private DBCon dbc=null;
        @Override
```

```
public List<Admin> findAdmins() {
    List<Admin> adminList=new ArrayList<Admin>();
    String sql="select * from admin order by id";
    try {
        dbc=new DBCon();
        ResultSet rs = dbc.doQueryRS(sql,new Object[]{});
        while(rs.next()){
            Admin admin=new Admin();
            admin.setId(rs.getInt("id"));
            admin.setUsername(rs.getString("username"));
            admin.setPassword(rs.getString("password"));
            adminList.add(admin);
        }
    } catch (SQLException e) {
        e.printStackTrace();
    }finally{
        dbc.close();
    }
    return adminList;
}
@Override
public Admin findByID(Integer id) {
    String sql="select * from admin where id=?";
    Admin admin=null;
    try {
        dbc=new DBCon();
        ResultSet rs = dbc.doQueryRS(sql, new Object[]{id});
        if(rs.next()){
            admin=new Admin();
            admin.setId(rs.getInt("id"));
            admin.setUsername(rs.getString("username"));
            admin.setPassword(rs.getString("password"));
        }
    } catch (SQLException e) {
        e.printStackTrace();
    }finally{
        dbc.close();
    }
    return admin;
}
@Override
public int addAdmin(Admin admin) {
    String sql="insert into admin(username,password) values(?,?)";
    int res=0;
    try {
        dbc=new DBCon();
        res= dbc.doUpdate(sql, new Object[]{admin.getUsername(),admin.getPassword()});
    } catch (Exception e) {
```

```
                e.printStackTrace();
            }finally{
                dbc.close();
            }
            return res;
    }
    @Override
    public int delAdmin(Integer id) {
            String sql="delete from admin where id=?";
            int res=0;
            try {
                dbc=new DBCon();
                res=dbc.doUpdate(sql, new Object[]{id});
            } catch (Exception e) {
                e.printStackTrace();
            }finally{
                dbc.close();
            }
            return res;

    }
    @Override
    public int updateAdmin(Admin admin) {
            String sql="update admin set username=?,password=? where id=?";
            int res=0;
            try {
                dbc=new DBCon();
                res= dbc.doUpdate(sql, new Object[]{admin.getUsername(),admin.getPassword(),admin. getId()});
            } catch (Exception e) {
                e.printStackTrace();
            }finally{
                dbc.close();
            }
            return res;
    }
}
```

（6）在 src 下的 cap.service 子包下新建 AdminService 接口，编辑后的代码如下：

```
package cap.service;
import java.util.List;
import cap.bean.Admin;
public interface AdminService {
    public List<Admin> findAdmins();
    public Admin findByID(Integer id);
    public int addAdmin(Admin admin);
    public int delAdmin(Integer id);
    public int updateAdmin(Admin admin);
}
```

（7）在 src 下的 cap.service.impl 子包中新建 AdminService 接口实现类 AdminServiceImpl.java，编辑后的代码如下：

```
package cap.service.impl;
import java.util.List;
import cap.bean.Admin;
import cap.dao.AdminDAO;
import cap.dao.impl.AdminDAOImpl;
import cap.service.AdminService;
public class AdminServiceImpl implements AdminService {
	private AdminDAO adminDAO=new AdminDAOImpl();
	@Override
	public List<Admin> findAdmins() {
		return adminDAO.findAdmins();
	}

	@Override
	public Admin findByID(Integer id) {
		return adminDAO.findByID(id);
	}
	@Override
	public int addAdmin(Admin admin) {
		return adminDAO.addAdmin(admin);
	}
	@Override
	public int delAdmin(Integer id) {
		return adminDAO.delAdmin(id);
	}
	@Override
	public int updateAdmin(Admin admin) {
		return adminDAO.updateAdmin(admin);
	}
}
```

（8）在 src 的 cap.action 下新建类 AdminAction.java，编辑后的代码如下：

```
package cap.action;
import java.util.List;
import cap.bean.Admin;
import cap.service.AdminService;
import cap.service.impl.AdminServiceImpl;
import com.opensymphony.xwork2.ActionSupport;
public class AdminAction extends ActionSupport {
	private AdminService adminService=new AdminServiceImpl();
	private List<Admin> adminList;
	private Integer id;
	private Admin admin;
	public String list()
	{
```

```
            adminList=adminService.findAdmins();
            return SUCCESS;
        }
        public String del()
        {
            int res=adminService.delAdmin(id);
            if(res>0)
                return SUCCESS;
            else
                return ERROR;
        }
        public String add()
        {
            int res=adminService.addAdmin(admin);
            if(res>0)
                return SUCCESS;
            else
                return ERROR;
        }
        public String edit()
        {
            admin=adminService.findByID(id);
            return SUCCESS;
        }
        public String update()
        {
            int res=adminService.updateAdmin(admin);
            if(res>0)
                return SUCCESS;
            else
                return ERROR;
        }
        //省略 getters 和 setters
}
```

（9）在 src 下新建 struts.xml，编辑后的代码如下：

```
<?xml version="1.0" encoding="UTF-8" ?>
<!DOCTYPE struts PUBLIC
        "-//Apache Software Foundation//DTD Struts Configuration 2.3//EN"
        "http://struts.apache.org/dtds/struts-2.3.dtd">
<struts>
        <package name="default" namespace="/" extends="struts-default">
            <action name="list" class="cap.action.AdminAction" method="list">
                <result name="success">/listAdmin.jsp</result>
            </action>
            <action name="del" class="cap.action.AdminAction" method="del">
                <result name="success" type="redirect">list</result>
                <result name="error">/error.jsp</result>
```

```
            </action>
            <action name="edit" class="cap.action.AdminAction" method="edit">
                <result name="success">/editAdmin.jsp</result>
            </action>
            <action name="update" class="cap.action.AdminAction" method="update">
                <result name="success" type="redirect">list</result>
                <result name="error">/error.jsp</result>
            </action>
            <action name="add" class="cap.action.AdminAction" method="add">
                <result name="success" type="redirect">list</result>
                <result name="error">/error.jsp</result>
            </action>
        </package>
    </struts>
```

（10）在 WebContent 下新建 index 页面，在 body 标签中添加下面的代码：

```
<a href="list">显示所有用户</a>
```

（11）在 WebContent 下新建 listAdmin.jsp，编辑后的页面代码如下：

```
<%@ page language="java" contentType="text/html; charset=UTF-8"
    pageEncoding="UTF-8"%>
<%@ taglib uri="/struts-tags" prefix="s"%>
<!DOCTYPE html PUBLIC "-//W3C//DTD HTML 4.01 Transitional//EN" "http://www.w3.org/TR/html4/
loose.dtd">
<html>
<head>
<meta http-equiv="Content-Type" content="text/html; charset=UTF-8">
<title>显示所有用户</title>
</head>
<body>
    <table>
        <s:iterator value="adminList" var="admin">
            <tr>
                <td><s:property value="#admin.id" /></td>
                <td><s:property value="#admin.username" /></td>
                <td><s:property value="#admin.password" /></td>
                <td><a href="del?id=<s:property value="#admin.id"/>">删除</a></td>
                <td><a href="edit?id=<s:property value="#admin.id"/>">编辑</a></td>
            </tr>
        </s:iterator>
    </table>
    <a href="addAdmin.jsp">添加用户</a>
</body>
</html>
```

（12）在 WebContent 下新建 editAdmin.jsp 页面，编辑后的代码如下：

```
<%@ page language="java" contentType="text/html; charset=UTF-8"
        pageEncoding="UTF-8"%>
<%@ taglib uri="/struts-tags" prefix="s"%>
<!DOCTYPE html PUBLIC "-//W3C//DTD HTML 4.01 Transitional//EN" "http://www.w3.org/TR/html4/loose.dtd">
<html>
<head>
<meta http-equiv="Content-Type" content="text/html; charset=UTF-8">
<title>更新</title>
</head>
<body>
        <s:form action="update" method="post">
                <s:hidden name="admin.id" />
                <s:textfield name="admin.username" label="用户名"></s:textfield>
                <s:textfield name="admin.password" label="密码"></s:textfield>
                <s:submit value="更新"></s:submit>
        </s:form>
</body>
</html>
```

（13）在 WebContent 下新建 addAdmin.jsp，编辑后的代码如下：

```
<%@ page language="java" contentType="text/html; charset=UTF-8"
        pageEncoding="UTF-8"%>
<%@ taglib    uri="/struts-tags" prefix="s" %>
<!DOCTYPE html PUBLIC "-//W3C//DTD HTML 4.01 Transitional//EN" "http://www.w3.org/TR/html4/loose.dtd">
<html>
<head>
<meta http-equiv="Content-Type" content="text/html; charset=UTF-8">
<title>Insert title here</title>
</head>
<body>
<s:form action="add" method="post">
<s:textfield name="admin.username" label="用户名"></s:textfield>
<s:textfield name="admin.password" label="密码"></s:textfield>
<s:submit value="添加"></s:submit>
</s:form>
</body>
</html>
```

（14）运行工程，在地址栏中输入 http://localhost:8080/struts9/list，会出现类似图 5-5 所示的运行结果。

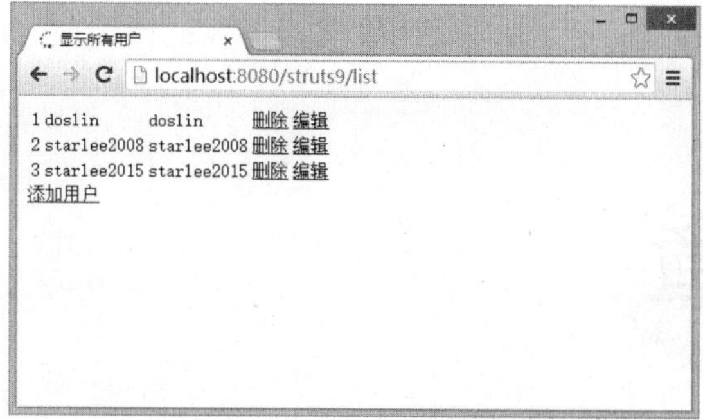

图 5-5　工程 struts9 的运行结果

第 6 章 Spring 框架与入门

为了解决企业应用程序开发复杂性日益增大的问题，人们提出并创建了一个轻量级的开源框架——Spring。它分层架构的理念深受技术开发人员的推崇，分层构架允许使用者使用哪一个组件，同时为 Java EE 应用程序开发提供集成的框架。

Spring 以控制反转（IoC）和面向切面的编程（Aspect Oriented Programming，AOP）为内核，提供了展示层 Spring MVC，持久层 Spring JDBC 以及业务层事务管理等众多的企业级应用技术，并能整合第三方的框架和内容，逐渐成为应用广泛的 Java EE 企业应用开源框架。

6.1 Spring 框架

Spring 框架的分层架构由 7 个定义良好的模块组成，Spring 模块构建在核心容器之上，核心容器定义了创建、配置和管理 Bean 的方式，如图 6-1 所示。

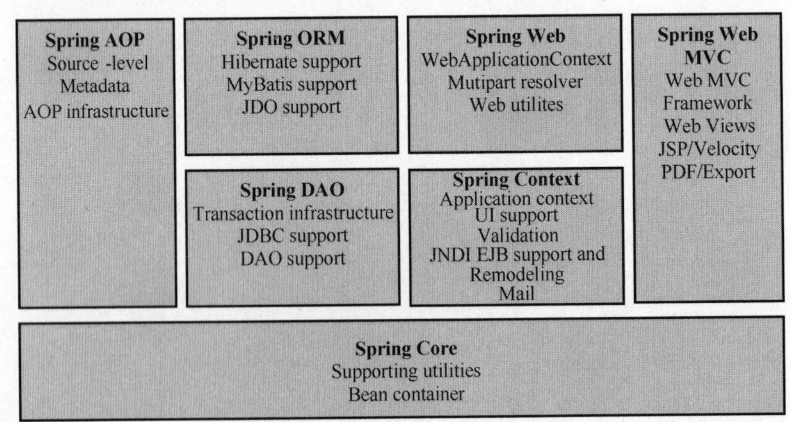

图 6-1 Spring 框架的 7 个模块

组成 Spring 框架的每个模块（或组件）既可以单独存在，也可以与其他一个或多个模块联合实现。每个模块的功能如下：

（1）核心容器：Spring 框架的基本功能由核心容器来实现，核心容器的主要组件是 BeanFactory，它是工厂模式的实现。BeanFactory 使用控制反转（IoC）模式将应用程序的配置和依赖性规范与实际的应用程序代码分开。

（2）Spring 上下文：Spring 上下文是一个配置文件，负责向 Spring 框架提供上下文信息。Spring 上下文包括企业服务，例如 EJB、JNDI、电子邮件、国际化、校验和调度等功能。

（3）Spring AOP：该模块通过配置管理特性直接将面向切面的编程功能集成到了 Spring 框架中。因此，可以很轻易地使 Spring 框架管理的任何对象支持 AOP。Spring AOP 模块为基于 Spring 的应用程序中的对象提供了事务管理服务。通过使用 Spring AOP，不用依靠 EJB 组件，就可以将声明性事务管理集成到应用程序中。

（4）Spring DAO：JDBC DAO 抽象层提供了有意义的异常层次结构，可用该结构来管理异常处理和不同数据库供给商抛出的错误消息。异常层次结构简化了错误处理，并且极大地降低了需要编写的异常代码数量（例如打开和关闭连接）。Spring DAO 的面向 JDBC 的异常遵从通用的 DAO 异常层次结构。

（5）Spring ORM：包含了对多个 ORM 框架的支持，从而方便进行对象关系映射，其中包括 JDO、Hibernate 和 MyBatis。所有这些都遵从 Spring 的通用事务和 DAO 异常层次结构。

（6）Spring Web 模块：Web 上下文模块建立在应用程序上下文模块之上，为基于 Web 的应用程序提供了上下文。所以，Spring 框架支持与 Jakarta Struts 的集成。Web 模块还简化了处理大部分请求以及将请求参数绑定到域对象的工作。

（7）Spring MVC 框架：Spring MVC 框架是一个全功能的构建 Web 应用程序的 MVC 实现。通过策略接口，MVC 框架变成为高度可配置的，MVC 容纳了大量视图技术，其中包括 JSP、Velocity、Tiles、iText 和 POI。

Spring 框架的功能可以用在任何 Java EE 服务器中，大多数功能也适用于不受管理的环境。Spring 的核心要点是：支持不绑定到特定 Java EE 服务的可重用业务和数据访问对象。这样的对象可以在不同 Java EE 环境（Web 或 EJB）、独立应用程序、测试环境之间重用。

6.2 Spring 开发入门

6.2.1 开发环境的搭建

搭建开发环境的具体步骤如下。

（1）打开 Eclipse 开发环境，首先需要安装 Spring 的插件 Spring Tool Suite，安装的方法和步骤详见附录 A-Eclipse 插件的安装。

（2）在 Eclipse 中选择“File→New”命令，新建 Java Project，如图 6-2 所示，单击“Next”按钮，在图 6-3 中给工程命名为 spring1，然后单击“Finish”按钮，工程创建完毕。

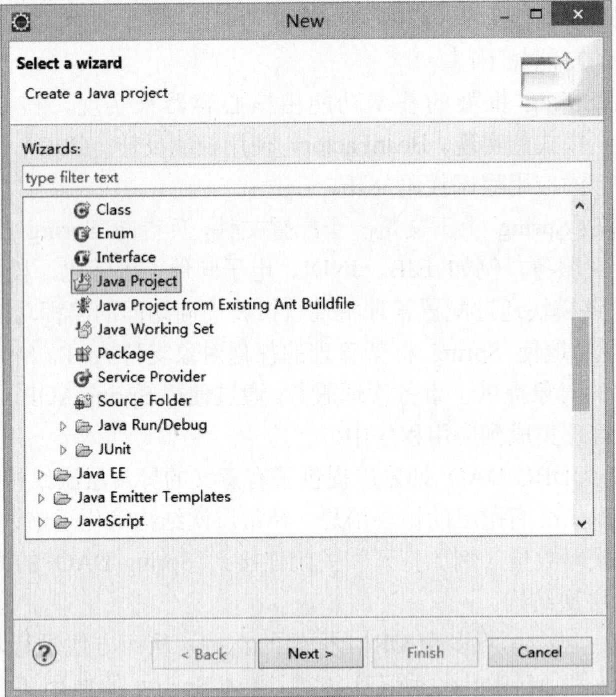

图 6-2　创建 Java 工程

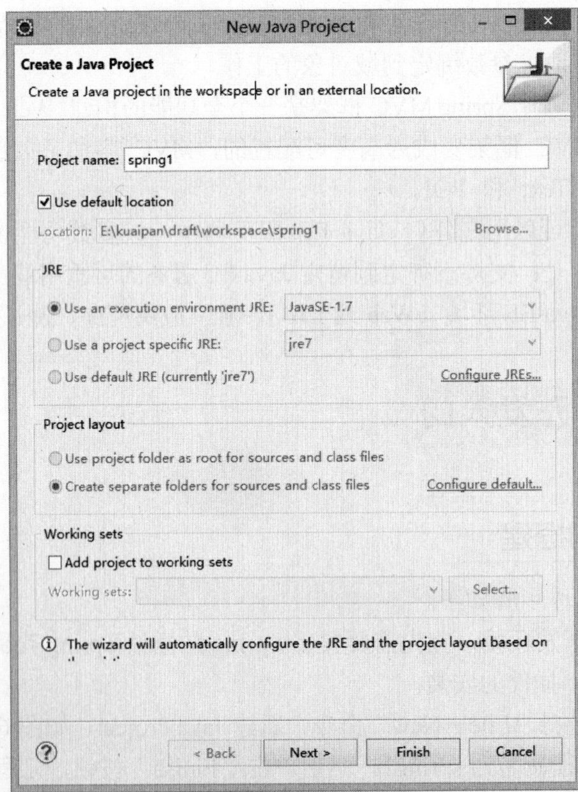

图 6-3　工程命名

（3）选中已经创建好的工程 spring1，右击，在打开的快捷菜单中选择“Properties”，出现“Properties for spring1”对话框，左边选中“Java Build Path”同时右边选中“Libraries”，如图 6-4 所示。

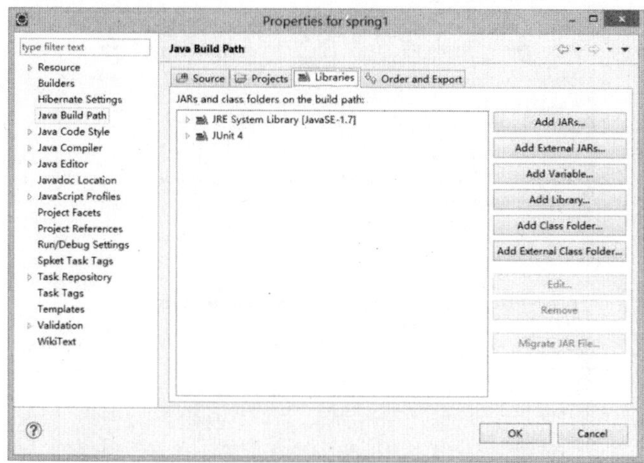

图 6-4 工程属性

（4）单击“Add External JARs”，出现如图 6-5 所示的“JAR Selection”对话框，到指定的目录选中如图 6-5 所示的全部的 Jars，然后单击“打开”按钮，接着单击“OK”按钮，对话框消失。

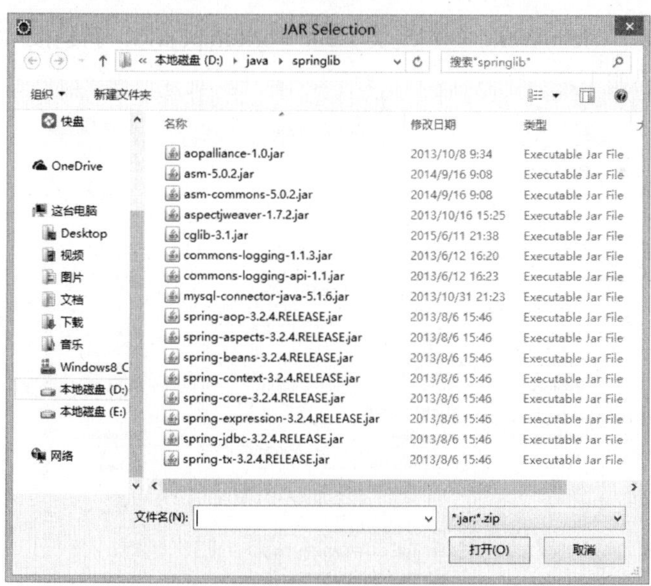

图 6-5 spring 开发所需的包

6.2.2 代码编写

（1）展开工程 spring1，选中 src 并右击，选择创建 HelloWorld 接口，出现创建接口的向导，如图 6-6 所示，在 Package 中输入“cap.servill”，Name 中输入“HelloWorld”，单击“Finish”

按钮，编辑代码如下。

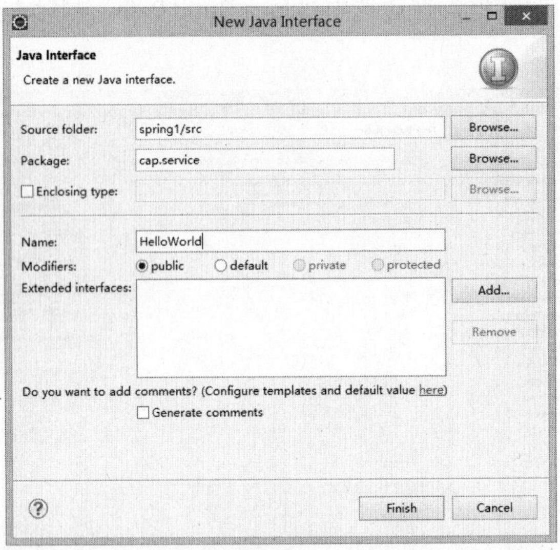

图 6-6　新建接口

```
package cap.service;
public interface HelloWorld {
    public String sayHi(String name);
}
```

（2）接着创建接口 HelloWorld 的实现类 HelloWorldImpl.java，新建接口实现类向导如图 6-7 所示，编辑后的代码如下。

图 6-7　新建接口实现类

```
package cap.service.impl;
import cap.service.HelloWorld;
public class HelloWorldImpl implements HelloWorld {
    @Override
    public String sayHi(String name) {
        return "欢迎您学习 Spring"+name;
    }
}
```

6.2.3 配置文件编写

（1）选中 src，在弹出的快捷菜单中选择“new→Other”命令，在对话框中依次选择“Spring→Spring Bean Configuration File”，如图 6-8 所示。在接下来的如图 6-9 所示的对话框中输入 Spring 配置文件的名称：applicatonContext.xml，接着单击“Next”按钮。出现如图 6-10 所示对话框，选择图 6-10 中所用到的 Spring 的组件的命名空间。

图 6-8 使用 Spring Tool Suite 创建配置文件

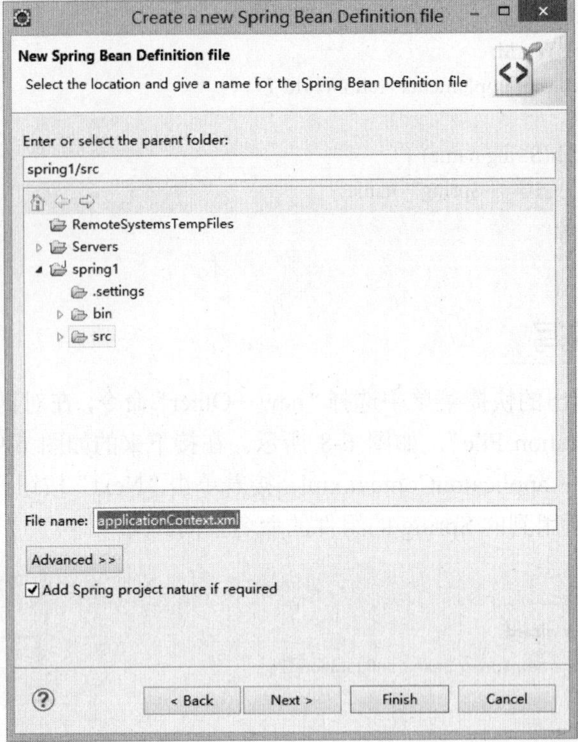

图 6-9　命名 Spring 配置文件

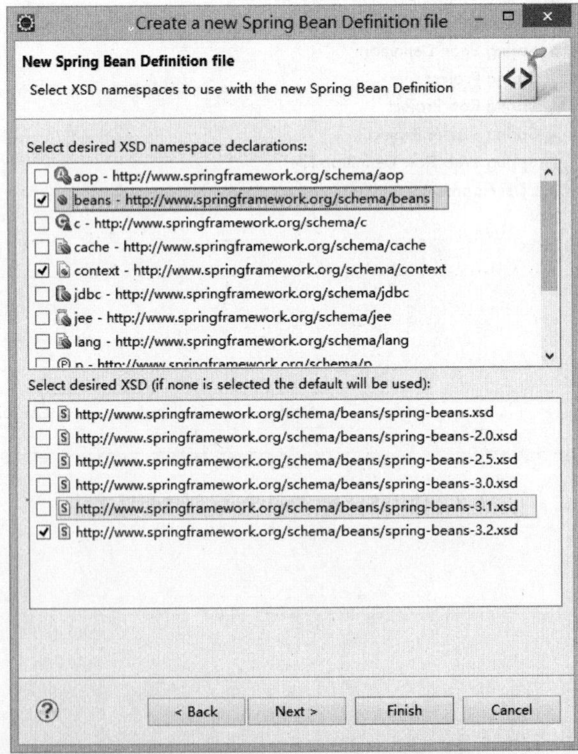

图 6-10　选择 Spring 组件的命名空间

（2）编辑 applictionContext.xml 的内容，代码如下。

```
<?xml version="1.0" encoding="UTF-8"?>
<beans xmlns="http://www.springframework.org/schema/beans"
        xmlns:xsi="http://www.w3.org/2001/XMLSchema-instance"
        xmlns:context="http://www.springframework.org/schema/context"
        xsi:schemaLocation="http://www.springframework.org/schema/beans    http://www.springframework.org/
schema/beans/spring-beans-3.2.xsd
            http://www.springframework.org/schema/context  http://www.springframework.org/schema/context/
spring-context-3.2.xsd">
      <bean id="helloWorld" class="cap.service.impl.HelloWorldImpl">
      </bean>
</beans>
```

代码解释：<bean>标签定义了一个 Bean,id 的属性值为 helloWorld，对应的类为 HelloWorldImpl。

6.2.4 测试类编写

（1）选中 src 并右击，选择创建 JUnit 测试类，如图 6-11 所示。单击“Next”按钮，在出现的创建 JUnit 测试类向导中做如图 6-12 所示的设置。

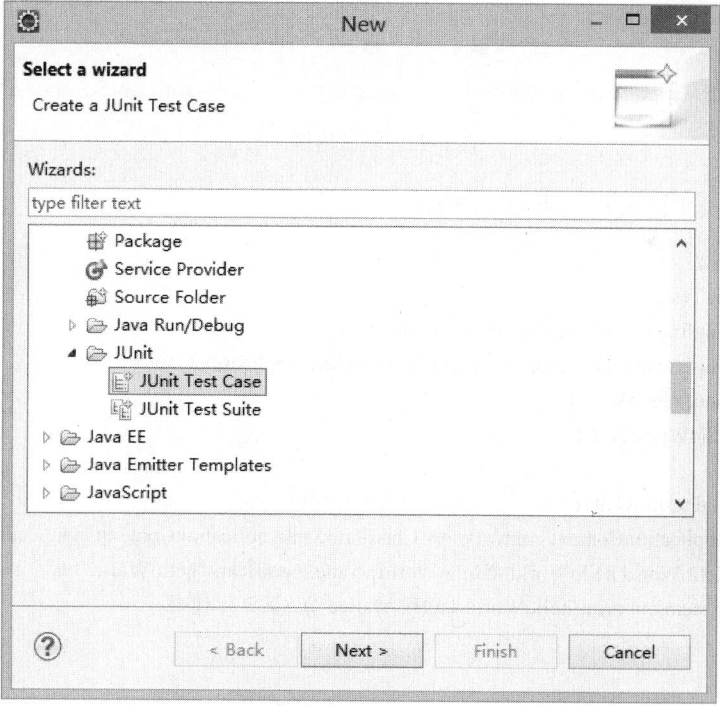

图 6-11 新建测试用例

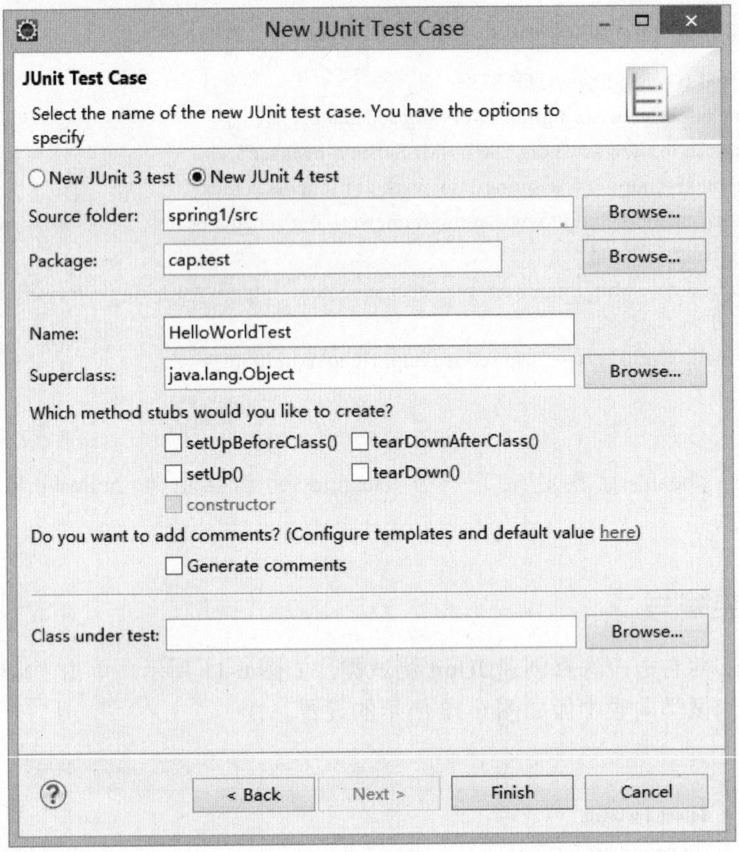

图 6-22　测试用例命名

（2）在测试类中的编辑代码如下所示。

```
package cap.test;
import org.junit.Test;
import org.springframework.context.ApplicationContext;
import org.springframework.context.support.ClassPathXmlApplicationContext;
import cap.service.HelloWorld;
public class HelloWorldTest {
        @Test
        public void testSayHi(){                    //第 1 行代码
                ApplicationContext context=new ClassPathXmlApplicationContext("applicationContext.xml");
                HelloWorld helloWorld=(HelloWorld) context.getBean("helloWorld");//第 2 行代码
                System.out.print(helloWorld.sayHi("starlee"));//第 3 行代码
        }
}
```

代码解释：第 1 行代码使用 ClassPathXmlApplicationContext 类加载 Spring 的配置文件 applicationContext.xml，初始化 Spring 的容器，接着使用 ApplicationContext 的 context 对象调用 getBean 获得 applicationContext.xml 中的 id="helloWorld"的 Bean 对象 helloWorld；第 3 行代码打印 helloWorld 对象中的 sayHi 方法添加返回的字符串。

（3）选中 HelloWorldTest.java 并右击，选择“Run As→Junit Test”，会在控制台上看见如

图 6-13 所示的内容。

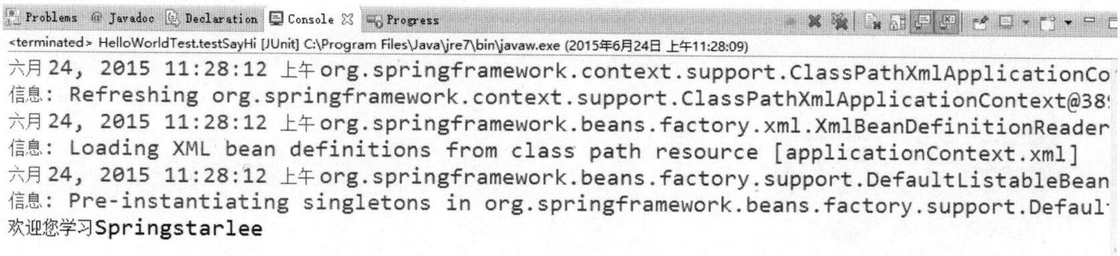
Problems
Javadoc
Declaration
Console
Progress
<terminated> HelloWorldTest.testSayHi [JUnit] C:\Program Files\Java\jre7\bin\javaw.exe (2015年6月24日 上午11:28:09)
六月 24, 2015 11:28:12 上午 org.springframework.context.support.ClassPathXmlApplicationCo
信息: Refreshing org.springframework.context.support.ClassPathXmlApplicationContext@38
六月 24, 2015 11:28:12 上午 org.springframework.beans.factory.xml.XmlBeanDefinitionReader
信息: Loading XML bean definitions from class path resource [applicationContext.xml]
六月 24, 2015 11:28:12 上午 org.springframework.beans.factory.support.DefaultListableBean
信息: Pre-instantiating singletons in org.springframework.beans.factory.support.Defaul
欢迎您学习Springstarlee

图 6-33　测试方法 sayHi 的运行结果

第7章 Spring IoC 容器

用户在使用 Spring 所提供的各种功能之前，必须在 Spring IoC 容器中装配好 Bean，并建立 Bean 和 Bean 之间的关联关系，Spring 提供了多种配置方式实现 Bean 的配置。

7.1 IoC 容器

IoC（Inversion of Control，控制反转）：其核心思想是反转资源获取的方向。传统的资源查找方式要求组件向容器发起请求查找资源，作为回应，容器适时地返回资源。而应用了 IoC 之后，则是容器主动地将资源推送给它所管理的组件，组件所要做的仅是选择一种合适的方式来接受资源。这种行为也被称为查找的被动形式。

依赖注入 DI（Dependency Injection）：是 IoC 的另一种表述方式，即组件以一些预先定义好的方式（例如：setter 方法）接受来自如容器的资源注入，相对于 IoC 而言，这种表述更直接。

典型的企业应用不会只由单一的对象（或 Bean）组成。毫无疑问，即使最简单的系统也需要多个对象一起来满足最终用户的需求。应用依赖注入（DI）原则后，代码会更加清晰。而且 Bean 无需担心对象之间的依赖关系（以及在何时何地指定这种依赖关系和依赖的实际类是什么）之后，也将更容易实现更高层次的低耦合。

在 Spring IoC 容器读取 Bean 配置创建 Bean 实例之前，必须对它进行实例化。只有在容器实例化后，才可以从 IoC 容器里获取 Bean 实例并使用。Bean 在 Spring IoC 容器中的初始化如图 7-1 所示。

Spring 提供了两种类型的 IoC 容器实现：

- BeanFactory：IoC 容器的基本实现。BeanFactory 是 Spring 框架的基础设施，面向 Spring 本身。
- ApplicationContext：提供了更多的高级特性。是 BeanFactory 的子接口。ApplicationContext 面向使用 Spring 框架的开发者，几乎所有的应用场合都直接使用 ApplicationContext 而非底层的 BeanFactory。

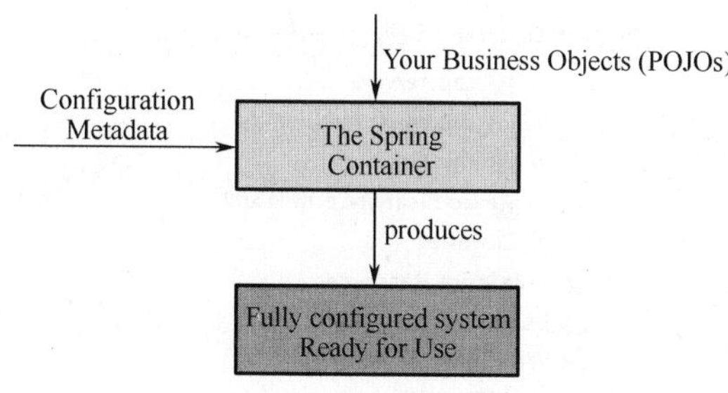

图 7-1 Spring IoC 容器

无论使用何种方式，配置文件是相同的。

7.2 BeanFactory

org.springframework.beans.factory.BeanFactory 是 Spring IoC 容器的实际代表者，IoC 容器负责容纳此前所描述的 Bean，并对 Bean 进行管理。

在 Spring 中，BeanFactory 是 IoC 容器的核心接口。它的职责包括：实例化、定位、配置应用程序中的对象及建立这些对象间的依赖。

Spring 提供了许多易用的 BeanFactory 实现，XmlBeanFactory 就是最常用的一个。该实现将以 XML 方式描述组成应用的对象以及对象间的依赖关系。XmlBeanFactory 类将获取此 XML 配置元数据，并用它来构建一个完全可配置的系统或应用。

BeanFactory 是通用工厂类，可以创建并管理各种类的对象。下面将通过一个具体的实例讲解 BeanFactory 实例化。

（1）在 Eclipse 中新建 Dynamic Web Project，工程名为 spring2，将 spring1 中 src 目录下的源码和配置文件复制到 spring2，其工程结构如图 7-2 所示。

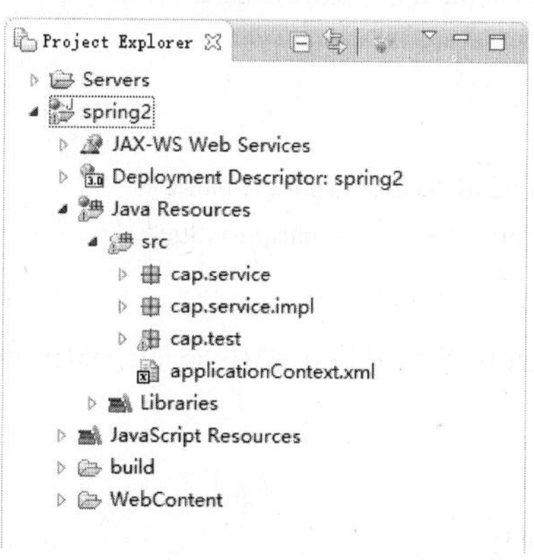

图 7-2 spring2 的工程结构图

（2）将开发 Spring 所需要的包复制到 WebContent 中 WEB-INF 下的 lib 目录，所需复制的 lib 如图 7-3 所示。

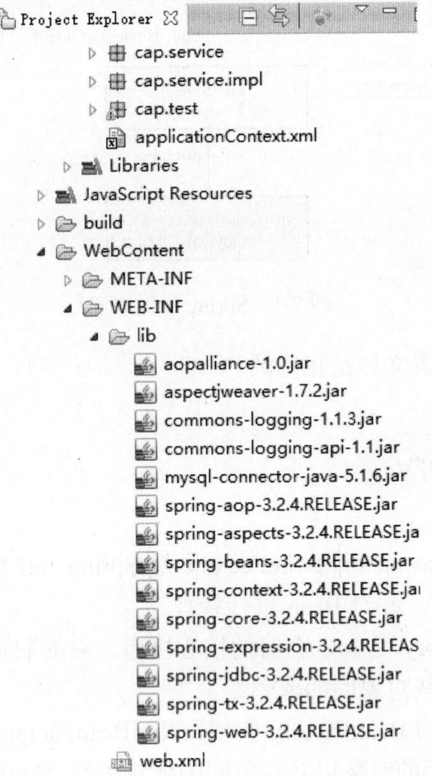

图 7-3　Spring 开发所需要的 Jar 包

（3）配置 Web.xml，在 web.xml 中添加下面的代码。

```
<listener>
        <listener-class>
        org.springframework.web.context.ContextLoaderListener
        </listener-class>
</listener>
<context-param>
        <param-name>contextConfigLocation</param-name>
        <param-value>applicationContext.xml</param-value>
</context-param>
```

（4）在 cap.test 子包中的 HelloWorldTest 类中添加下面的测试方法，最后的运行结果和上一节的运行结果相同。

```
@Test
    public void xmlBeanSayHi(){
        Resource resource = new ClassPathResource("applicationContext.xml");//第 1 行代码
        BeanFactory factory = new XmlBeanFactory(resource); //第 2 行代码
        HelloWorld helloWorld = (HelloWorld) factory.getBean("helloWorld");
        System.out.print(helloWorld.sayHi("starlee"));
    }
```

代码解释：第 1 行代码 ClassPathResource 加载 applicationContext.xml 中的资源，第 2 行代

码使用 XmlBeanFactory 类初始化 Spring 的容器。余下的代码和前面一节相似。

Spring IoC 容器将管理一个或多个 Bean，这些 Bean 将通过配置文件中的 Bean 定义创建在 applicationContext.xml 中。在本例中 Bean 的创建代码如下。

```
<bean id="helloWorld" class="cap.service.impl.HelloWorldImpl">  </bean>
```

7.3 ApplicationContext

Spring 通过一个配置文件描述 Bean 与 Bean 之间的依赖关系，利用 Java 语言的反射功能实例化 Bean 并建立 Bean 之间的依赖关系。Spring 的 IoC 容器在完成这些底层工作的基础上，还提供了 Bean 实例缓存、生命周期管理、Bean 实例代理、事件发布、资源装载等高级服务。

Bean 工厂是 Spring 框架最核心的接口，它提供了高级 IoC 的配置机制。BeanFactory 可以管理不同类型的 Java 对象，应用上下文 ApplicationContext 建立在 BeanFactory 基础之上，提供了更多面向应用的功能，它还提供了国际化支持和框架事件体系，更易于创建实际应用。一般称 BeanFactory 为 IoC 容器，称 ApplicationContext 为应用上下文。

BeanFactory 是 Spring 框架的基础设施，面向 Spring 本身。ApplicationContext 面向 Spring 框架的开发者，几乎所有的应用场合都直接使用 ApplicationContext 而非底层的 BeanFactory。

以下介绍 ApplicationContext 的四种主要实现类。

- ClassPathXmlApplicationContext：从类路径下加载配置文件。
- FileSystemXmlApplicationContext：从文件系统中加载配置文件。
- ApplicationContext：在初始化上下文时就实例化所有单例的 Bean。
- WebApplicationContext：是专门为 WEB 应用而准备的，它允许从相对于 WEB 根目录的路径中完成初始化工作。

7.3.1 获取 Bean

在 Spring 的容器中调用 ApplicationContext 的 getBean()方法，就可以获得实例化的 Bean 对象，getBean()方法的重载方式如表 7-1 所示。

表 7-1　getBean()方法的重载

方 法 名	功 能 说 明
Object getBean(String name) throws BeansException	根据 Bean 的 id 获取 Bean 实例，需强制类型转化
<T> T getBean(Class<T> requiredType) throws BeansException	根据 Bean 的类型获取 Bean 实例
<T> T getBean(String name, Class<T> requiredType)	该方法重载了 getBean(String name)方法，根据 Bean 的 id 和类型获取 Bean 实例
Object getBean(String name, Object... args)	可变参数主要用来指定是否显示使用静态工厂方法创建一个原型(prototype)Bean

7.3.2 ApplicationContext 实例化容器

Spring 中使用 ApplicationContext 实例化容器有 3 种，XML 是最常见的应用系统配置源。Spring 中的几种容器都支持使用 XML 装配 Bean，包括 ClassPathXmlApplicationContext、FileSystemXmlApplicationContext 和 XmlWebApplicationContext。

- ClassPathXmlApplicationContext：编译路径，使用规则如下。

```
ApplicationContext factory=new ClassPathXmlApplicationContext("appcontext.xml");
```

- FileSystemXmlApplicationContext：文件系统的路径，使用规则如下。

```
ApplicationContext factory=new FileSystemXmlApplicationContext("src/appcontext.xml");
```

- XmlWebApplicationContext：Web 工程，使用规则如下。

```
ServletContext servletContext = request.getSession().getServletContext();
ApplicationContext ctx = webApplicationContextUtils.getWebApplicationContext(servletContext );
```

注：其中 FileSystemXmlApplicationContext 和 ClassPathXmlApplicationContext 与 BeanFactory 的 xml 文件定位方式一样是基于路径的。

下面将通过一个具体的实例讲解通过 ApplicationContext 实例化 Spring 的容器。

（1）继续在工程 spring 中，在 cap.test 子包中的 HelloWorldTest 类中添加下面的测试方法，最后的运行结果和第 6 章的运行结果相同。

```
@Test
    public void fileXMlSayHi(){         //第 1 行代码
        ApplicationContext factory=new FileSystemXmlApplicationContext("src/applicationContext.xml");
        HelloWorld helloWorld = (HelloWorld) factory.getBean("helloWorld");
        System.out.print(helloWorld.sayHi("starlee"));
    }
```

代码解释：第 1 行代码通过文件系统路径加载 Spring 的配置文件 applicationContext.xml，用于初始化 Spring 容器。

（2）采用 XmlWebApplicationContext 记载 Spring 的配置资源，在 WebContent 下新建 spring.jsp 页面，编辑后的代码如下。

```
<%@page import="cap.service.HelloWorld"%>
<%@page import="org.springframework.web.context.support.WebApplicationContextUtils"%>
<%@page import="org.springframework.context.ApplicationContext"%>
<%@ page contentType="text/html; charset=UTF-8" pageEncoding="UTF-8"%>
<!DOCTYPE html PUBLIC "-//W3C//DTD HTML 4.01 Transitional//EN" "http://www.w3.org/TR/html4/loose.dtd">
<html>
<head>
<meta http-equiv="Content-Type" content="text/html; charset=UTF-8">
<title>采用 XmlWebApplicationContext 加载 spring 的配置文件</title>
</head>
<body>
<%
```

```
//第 1 行代码
ServletContext servletContext = request.getSession().getServletContext();
//第 2 行代码
ApplicationContext ctx = WebApplicationContextUtils.getWebApplicationContext(servletContext );
HelloWorld helloword=(HelloWorld)ctx.getBean("helloWorld");
String str=helloword.sayHi("starlee2008");
%>
<%=str %>
</body>
</html>
```

代码解释：第 1 行代码通过 request 内置对象获得 ServletContext 的对象 servletContext，第 2 行代码通过 WebApplicationContextUtils 工具类的静态 getWebApplicationContext 方法初始化 Spring 容器。

（3）页面运行的结果如图 7-4 所示。

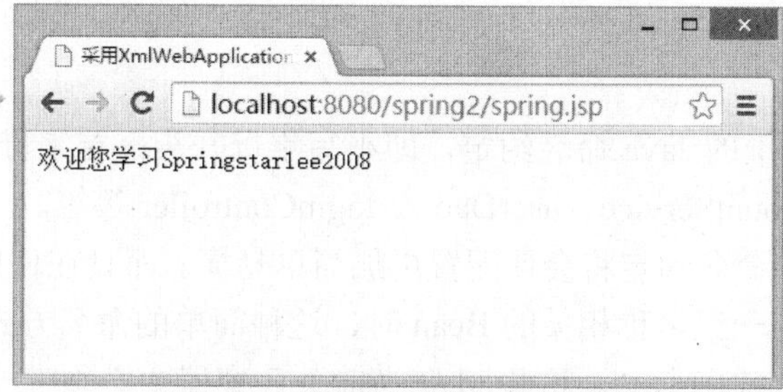

图 7-4　spring2 运行结果

第8章 Spring Bean

Bean 的命名采用标准的 Java 命名约定，即小写字母开头，首字母大写间隔的命名方式。如 accountManager、accountService、userDao 及 loginController 等等。

对 Bean 采用统一的命名约定将会使配置更加简单易懂。而且在使用 Spring AOP 时，如果要发通知（advice）给与一组名称相关的 Bean 时，这种简单的命名方式将会令你事半功倍。

每个 Bean 都有一个或多个 id。这些 id 在当前 IoC 容器中必须唯一。如果一个 Bean 有多个 id，那么其他的 id 在本质上将被认为是别名。

Bean 配置形式主要有两种方式：基于 XML 文件的方式和基于注解的方式。下面先讲解基于 XML 文件的方式配置 Bean。

8.1 基于 XML 文件的方式配置 Bean

8.1.1 Bean 的配置方式

Bean 的配置方式有：使用类构造器直接实例化、使用静态工厂的方法实例化和使用实例工厂方法实例化。

通过类构造器实现配置 Bean，主要是在 Spring 的配置文件 applicationContext.xml 中通过 <bean>节点来配置。<bean>元素的主要属性见表 8-1。

表 8-1 <bean>元素的主要属性

属 性 名	功 能 说 明
id	指定 Bean 的唯一标识名，通过 id 值可以获取 Bean。id 属性值在 IoC 容器中必须是唯一的；若没有指定 id，Spring 自动将全限定性类名作为 Bean 的名字
class	指定 Bean 的全类名（包名＋类名），通过反射的方式在 IoC 容器中创建 Bean，要求 Bean 中必须有默认无参构造器。只有子类 Bean 不用定义该属性

续表

属 性 名	功 能 说 明
name	用来为 id 创建一个或多个别名。它可以是任意的字母符合。多个别名之间用逗号或空格分开
parent	表示继承的父类。如果有很多继承同一个父类的 Bean，那么在配置文件中实例那些 Bean 时候可以省略掉父类已经注入的属性。Bean 定义继承父 Bean 定义，它可以覆盖父 Bean 的一些值，或者它需要的值。那么在配置文件中实例那些 Bean 的时候可以省略掉父类已经注入的属性。
abstract	默认为“false”。用来定义 Bean 是否为抽象 Bean。它表示这个 Bean 将不会被实例化，一般用于父类 Bean，因为父类 Bean 主要是供子类 Bean 继承使用。
lazy-init	默认为“default”。用来定义这个 Bean 是否实现懒加载。如果为“true”，它将在 BeanFactory 启动时初始化所有的 Singleton Bean。反之，如果为“false”,它只在 Bean 请求时才开始创建 Singleton Bean。
depends-on	依赖对象。这个 Bean 在初始化时依赖的对象，这个对象会在这个 Bean 初始化之前创建。
init-method	用来定义 Bean 的初始化方法，它会在 Bean 组装之后调用。它必须是一个无参数的方法。
destroy-method	用来定义 Bean 的销毁方法，它在 BeanFactory 关闭时调用。同样，它也必须是一个无参数的方法。它只能应用于 singleton Bean。
factory-method	定义创建该 Bean 对象的工厂方法。它用于下面的“factory-bean”，表示这个 Bean 是通过工厂方法创建。此时，“class”属性失效。
factory-bean	定义创建该 Bean 对象的工厂类。如果使用了“factory-bean”，则“class”属性失效。
autowire	自动装配，默认为 default。它定义了 Bean 的自动装载方式。 no：不使用自动装配功能。 byName：通过 Bean 的属性名实现自动装配。 byType：通过 Bean 的类型实现自动装配。 constructor：类似于 byType，但它是用于构造函数的参数的自动组装。不推荐使用。 autodetect：通过 Bean 类的反省机制（introspection）决定是使用 constructor 还是使用 byType。
autowire-candidate	默认“default”。
primary	默认“true”。
scope	设置 Bean 的作用域，默认值为“singleton”。有 5 种类型 singleton、prototype、request、session、globalSession。

由于使用类构造器直接实例化在前面的案例中已经讲解过，所以下面将通过一个具体的案例来讲解 Spring Bean 的其余两种管理方式。

（1）打开 Eclipse 开发环境，选择 Package Explorer，选中工程 spring1，在弹出的快捷菜单中选择“copy”，然后在 Package Explorer 的空白处右击，在打开的快捷菜单中选择“paste”，然后在弹出如图 8-1 所示窗口中的 Project name 输入框中输入工程名：spring3。

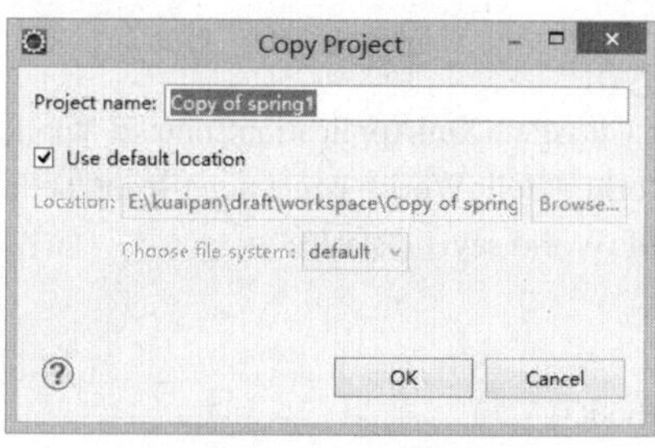

图 8-1 复制拷贝

（2）使用静态工厂的方法实例化。在 src 下创建 cap.service 包，在该包下创建 HelloWorldBeanFacotry 类，编辑的代码如下。

```
package cap.service.impl;
import cap.service.HelloWorld;
public class HelloWorldBeanFactory implements HelloWorld {
        public static    HelloWorld createHelloWorldBean()
        {
                return new HelloWorldImpl();
        }
        public HelloWorld createHelloWorldBean2()
        {
                return new HelloWorldImpl();
        }
        @Override
        public String sayHi(String name) {
                return "欢迎您学习 Spring"+name;
        }
}
```

（3）对应的在 applicationContext.xml 中的配置如下。

```
    <bean id="helloWorld1" class="cap.service.impl.HelloWorldBeanFactory" factory-method=
"createHelloWorldBean">
    </bean>
```

代码解释：在<bean>标签中指定 factory-method="createHelloWorldBean"，使用静态工厂方法实例化 Bean 对象 helloWorld1。

（4）使用实例工厂方法实例化，在 applicationContext.xml 中添加下面的 Bean 配置文件。

```
<bean id="helloWorldFactory" class="cdap.impl.HelloWorldBeanFactory">
</bean>
<bean id="helloWorld2" factory-bean="helloWorldFactory" factory-method="createHelloWorldBean2">
</bean>
```

代码解释：第一个<bean>标签指定工厂 Bean 为 helloWorldFactory，第 2 个<bean>中的 factory-bean 指定工厂 Bean 为 helloWorldFactory，factory-method 属性指定工厂方法为 createHelloWorldBean2

（5）在 cap.test 下的 HelloWorldTest.java 类中添加下面两个新测试方法。

```
        @Test
        public void testStaticFactory(){
    ApplicationContext context=new ClassPathXmlApplicationContext("applicationContext.xml");
                HelloWorld helloWorld=(HelloWorld) context.getBean("helloWorld1");
                System.out.print(helloWorld.sayHi("starlee"));
        }
        @Test
        public void testFactoryMethod(){
                ApplicationContext context=new ClassPathXmlApplicationContext("applicationContext.xml");
```

```
            HelloWorld helloWorld=(HelloWorld) context.getBean("helloWorld2");
            System.out.print(helloWorld.sayHi("starlee"));
        }
```

代码解释: testStaticFactory 方法用于测试静态工厂实例化 Bean 对象，testFactoryMethod 方法用于测试工厂方法实例化 Bean 对象。

8.1.2 Bean 的作用域

Bean 的作用域有：singleton、prototype、request、session、global session 五种，其中后三者是对应于 Web 环境变量下的作用域。

在 Spring 中，可以在<bean>元素的 scope 属性里设置 Bean 的作用域。在默认情况下，Spring 只为每个在 IoC 容器里声明的 Bean 创建唯一一个实例，整个 IoC 容器范围内都能共享该实例。所有后续的 getBean()调用和 Bean 引用都将返回这个唯一的 Bean 实例。该作用域被称为 singleton，它是所有 Bean 的默认作用域。

Bean 作用域以及描述如表 8-2 所示。

表 8-2 Bean 的作用域

作 用 域	描 述
singleton	在每个 Spring IoC 容器中一个 Bean 定义对应一个对象实例
prototype	一个 Bean 定义对应多个对象实例
request	在一次 HTTP 请求中，一个 Bean 定义对应一个实例；即每次 HTTP 请求将会有各自的 Bean 实例，它们依据某个 Bean 定义创建而成。该作用域仅在基于 Web 的 Spring ApplicationContext 情形下有效
session	在一个 HTTP Session 中，一个 Bean 定义对应一个实例。该作用域仅在基于 Web 的 Spring ApplicationContext 情形下有效
global session	在一个全局的 HTTP Session 中，一个 Bean 定义对应一个实例。典型情况下，仅在使用 portlet context 的时候有效。该作用域仅在基于 Web 的 Spring ApplicationContext 情形下有效

下面将通过一个具体的实例讲解 Bean 的作用域。

（1）使用工程 spring3，在 src 的 cap.test 中的 HelloWorldTest 测试类中添加下面的测试方法。

```
    @Test
        public void testBeanSingletonScope(){
            HelloWorld helloWorld1=(HelloWorld) context.getBean("helloWorld");
            HelloWorld helloWorld2=(HelloWorld) context.getBean("helloWorld");
            if(helloWorld1.equals(helloWorld2)){
                System.out.println("Spring Bean 的作用域默认为 singleton，所以 helloWorld1 和
helloWorld2 是同一个对象");
            }
        }
```

（2）运行 testBeanSingletonScope 方法之后，会在控制台下打印出“Spring Bean 的作用域默认为 singleton，所以 helloWorld1 和 helloWorld2 是同一个对象”。这是因为 Spring 的 Bean 的默认作用域为 singleton，在 Spring 的容器中创建唯一的实例，所以获得的都是同一个实例。

（3）在 src 的 Spring 配置文件 applicationContext.xml 中添加下面的 Bean 的配置，此处 Bean

的作用域为 prototype。

```
<bean id="helloWorld3" class="cap.service.impl.HelloWorldImpl" scope="prototype">
</bean>
```

（4）在 src 的 cap.test 中的 HelloWorldTest 测试类中添加下面的测试方法。

```
@Test
    public void testBeanPrototypeScope(){
        HelloWorld helloWorld1=(HelloWorld) context.getBean("helloWorld3");
        HelloWorld helloWorld2=(HelloWorld) context.getBean("helloWorld3");
        if(helloWorld1.equals(helloWorld2)){
System.out.println("helloWorld1 和 helloWorld2 是同一个对象");
        } else {
System.out.println("helloWorld1 和 helloWorld2 是不同的对象");
        }
    }
```

（5）运行 testBeanPrototypeScope 方法之后，会在控制台下打印出“helloWorld1 和 helloWorld2 是不同的对象”。这是因为 Spring 的 Bean 的作用域为 prototype，在 Spring 的容器中维护了不同的对象实例，所以获得的不是同一个实例。

8.1.3 依赖注入

Spring 主要有两种依赖注入的方式：属性注入和构造器注入。

1. 属性注入

属性注入：通过属性（setter）方法注入 Bean 的属性值或依赖的对象。属性注入使用<property>元素，使用 name 属性指定 Bean 的属性名称，value 属性或<value>子节点指定属性值。<property>元素的主要属性如表 8-3 所示。

表 8-3 <property>元素的属性

属 性 名	功 能 说 明
name	指定 Bean 的属性名称。必须属性
value	指定属性值（注入的常量值）
ref	指定属性引用（引用其他 Bean）

通过调用无参数构造器或无参数静态（static）工厂方法实例化 Bean 之后，调用该 Bean 的 setter 方法，即可实现基于 setter 的依赖注入。

下面的实例将展示使用属性(setter)注入依赖。

（1）创建 Java 工程 spring4，在 src 的 cap.bean 下创建 Admin.java 类，编辑后的代码如下。

```
package cap.bean;
public class Admin {
    private int id;
    private String username;
    private String password;
    public Admin() {}
```

```
    public Admin(int id, String username, String password) {
        this.id = id;
        this.username = username;
        this.password = password;
    }
//省略 getters 和 setters
}
```

（2）在 scr 下新建 applicationContext.xml，编辑后的代码如下。

```
<?xml version="1.0" encoding="UTF-8"?>
<beans xmlns="http://www.springframework.org/schema/beans"
       xmlns:xsi="http://www.w3.org/2001/XMLSchema-instance"
       xsi:schemaLocation="http://www.springframework.org/schema/beans
           http://www.springframework.org/schema/beans/spring-beans.xsd">
  <bean id="admin1" class="cap.bean.Admin">
          <property name="id" value="1"/>
                <property name="username" value="starlee2008"/>
                <property name="password" value="starlee2008"/>
  </bean>
</beans>
```

代码解释：<bean>标签中的<property>标签根据属性名注入具体的字面值。Spring 会根据属性的类型自动转化注入到对象 admin1 的属性中。

2. 构造方法注入

通过构造方法注入 Bean 的属性值或依赖的对象，它保证了 Bean 实例在实例化后就可以使用。<constructor-arg>元素里主要的属性如表 8-4 所示。

表 8-4 <constructor-arg>元素的属性

属 性 名	功 能 说 明
name	表示需要匹配的参数名字
index	表示参数索引，从 0 开始，即第一个参数索引为 0
type	表示需要匹配的参数类型，可以是基本类型也可以是其他类型（必须全限定类名），如“int”、“java.lang.String”
value	用来指定注入的常量值
ref	用来引用其他 Bean

默认按<bean>元素中<constructor-arg>的个数和顺序匹配构造器。

使用<constructor-arg>注入主要有三种方式：指定的 index 属性值作为索引匹配入参；指定 type 属性值作为类型匹配入参；按指定的 name 属性值作为参数名匹配入参。

下面将通过一个具体实例讲解通过构造器方法注入的使用。

（1）使用构造器参数来注入依赖关系有两种方式，根据参数索引和参数名注入，在 src 下的 applicationContext.xml 中添加，下面分别是两种不同配置方式。参数索引注入方式如下：

```
<bean id="admin2" class="cap.bean.Admin">
   <constructor-arg index="0" value="1"/>
   <constructor-arg index="1" value="cdap"/>
```

```
    <constructor-arg index="2" value="cdap"/>
</bean>
```

参数名注入方式如下。

```
<bean id="admin3" class="cap.bean.Admin">
        <constructor-arg name="id" value="1"/>
        <constructor-arg name="username" value="cdavtc"/>
        <constructor-arg name="password" value="cdavtc"/>
</bean>
```

（3）在 cap.test 子包中新建 JUnit 测试类 AdminTest.java，编辑后的代码如下。

```
package cdap.test;
import org.junit.Test;
import org.springframework.context.ApplicationContext;
import org.springframework.context.support.ClassPathXmlApplicationContext;
import cdap.bean.Admin;
public class AdminTest {
    private ApplicationContext ctx = new ClassPathXmlApplicationContext("applicationContext.xml");
    @Test
    public void testAdmin1(){
        Admin admin=(Admin) ctx.getBean("admin1");
        System.out.println(admin.getId()+":"+admin.getUsername()+":"+admin.getPassword());
    }
    @Test
    public void testAdmin2(){
        Admin admin=(Admin) ctx.getBean("admin2");
        System.out.println(admin.getId()+":"+admin.getUsername()+":"+admin.getPassword());
    }
    @Test
    public void testAdmin3(){
        Admin admin=(Admin) ctx.getBean("admin3");
        System.out.println(admin.getId()+":"+admin.getUsername()+":"+admin.getPassword());
    }
}
```

代码解释：testAdmin1 方法通过 ApplicationContext 对象 ctx 获得配置文件 applicationContext.xml 中<bean>标签中的 id="admin1"的对象，然后打印 admin1 对象的属性。其余的两个测试方法功能相似，不再做具体的解释。

由于大量的构造器参数可能使程序变得笨拙，特别是当某些属性是可选的时候。因此通常情况下，Spring 开发团队提倡使用 setter 参数注入。

8.1.4 注入属性值的类型

Spring 注入属性的类型包括字面值，注入 Bean 和注入集合等，在本节中主要讲解字面值和 Bean 的注入。

1. 字面值

字面值：可用字符串表示的值，可以通过<value>元素标签或 value 属性进行注入。基本数据类型及其封装类、String 等类型都可以采取字面值注入的方式。若字面值中包含特殊字符，可以使用<![CDATA[]]>把字面值包裹起来。下面是从构造器方法注入工程中截取的代码片段，实现的功能就是注入字面值。

```
<bean id="admin2" class="cdap.bean.Admin">
        <constructor-arg index="0" value="1"/>
        <constructor-arg index="1" value="cdap"/>
        <constructor-arg index="2" value="cdap"/>
</bean>
```

2. 引用其他 Bean

组成应用程序的 Bean 经常需要相互协作以完成应用程序的功能。要使 Bean 能够相互访问，就必须在 Bean 配置文件中指定对 Bean 的引用。在 Bean 的配置文件中，可以通过<ref>元素或 ref 属性为 Bean 的属性指定对 Bean 的引用。

下面将通过一个具体的实例讲解引用其他 Bean。

（1）在工程 spring4 的 src 下的 cap.service 子包中添加下面的接口。

```
package cap.service;
public interface AdminService {
    public String sayHello();
}
```

（2）在 src 的 cap.service.impl 子包中新建 AdminService 接口的实现类 AdminServiceImpl.java，编辑后的代码如下。

```
package cap.service.impl;
import cap.bean.Admin;
import cap.service.AdminService;
public class AdminServiceImpl implements AdminService {
    private Admin admin;
    public AdminServiceImpl() {
    }
    public AdminServiceImpl(Admin admin) {
        this.admin = admin;
    }
    public Admin getAdmin() {
        return admin;
    }
    public void setAdmin(Admin admin) {
        this.admin = admin;
    }
    public String sayHello(){
        return "你好：" +admin.getUsername();
    }
}
```

（3）在 src 的 Spring 配置文件 applicationContext.xml 中添加下面的代码。

```
<bean id="admin" class="cap.bean.Admin">
        <property name="id" value="1"/>
        <property name="username" value="starlee2008"/>
        <property name="password" value="starlee2008"/>
</bean>
<bean id="adminService" class="cap.service.impl.AdminServiceImpl">
        <property name="admin" ref="admin">
        </property>
</bean>
<bean id="adminService1" class="cap.service.impl.AdminServiceImpl">
        <property name="admin">
                <ref bean="admin"/>
        </property>
</bean>
```

代码解释：第一个 Bean 标签创建 id="admin"的对象，第二个<bean>标签创建 id="adminService"的对象，此对象 adminService 有一个属性 admin，需要引用第 1 个<bean>标签创建的 admin 对象。第 3 个<bean>标签实现的作用和第 2 个<bean>标签相似，只是引用对象的使用方式不同。

（4）在 cap.test 中新建 Junit 测试类 AdminServiceTest.java，添加下面的测试方法。

```
@Test
public void testSayHello(){
        AdminService adminService=(AdminService) ctx.getBean("adminService");
        System.out.println(adminService.sayHello());
        }
@Test
public void testSayHello1(){
        AdminService adminService=(AdminService) ctx.getBean("adminService1");
        System.out.println(adminService.sayHello());
}
```

3. 内部 Bean

当 Bean 实例仅仅给一个特定的属性使用时，可以将其声明为内部 Bean。内部 Bean 声明直接包含在<property>或<constructor-arg>元素里，内部 Bean 作用范围仅限在包含的 Bean 范围有效，不能使用在其他地方。

（1）在工程 spring4 中编辑，在 src 的 Spring 配置文件 applicationContext.xml 中添加下面的配置代码。

```
<bean id="adminService2" class="cap.service.impl.AdminServiceImpl">
        <property name="admin">
                <bean id="admin" class="cap.bean.Admin">
                        <property name="id" value="1"/>
                        <property name="username" value="starlee2008"/>
                        <property name="password" value="starlee2008"/>
        </bean>
```

```
        </property>
    </bean>
```

代码解释：<bean>标签 id="admin"的作用范围仅在<bean>标签 id=adminService2 的范围有效。

（2）在 cap.test 中新建 Junit 测试类 AdminServiceTest.java，添加下面的测试方法。

```
@Test
public void testSayHello2(){
    AdminService adminService=(AdminService) ctx.getBean("adminService2");
    System.out.println(adminService.sayHello());
}
```

8.2 基于注解的方式配置 Bean

目前比较流行的基于注解的配置方式包括基于注解配置 Bean 和基于注解来装配 Bean 的属性。

Spring 容器默认禁止注解装配。在使用基于注解的自动装配前，需要在 Spring 配置中启用它。最简单的方法是使用 Spring 的 context 命名空间配置中的<context:annotation-config/>元素。

8.2.1 组件扫描

组件扫描(component scanning)：Spring 能够从 classpath 中自动扫描、侦测和实例化具有特定注解的组件。默认情况下，<context:component-scan>查找使用构造型（sterotype）注解所标注的类，这些特定的注解如下：

- @Component：基本注解，标识了一个受 Spring 管理的组件（Bean），可以作用在任何层次，在不能确定是哪一个层的时候使用。
- @Respository：标识持久层组件，就是常说的 DAO 层。
- @Service：标识服务层（业务层）组件，就是常说的 service 层。
- @Controller：标识表现层（控制层）组件，就是常说的 Controller 或 action 层。

这四个注解都标识了一个受 Spring 管理的组件，Spring 目前无法区分它们，可以混用，但推荐按照其含义使用注解。

对于扫描到的组件，Spring 有默认的命名策略：使用非限定类名（第一个字母小写）。例如，cap.service.AdminService→adminService。也可以在注解中通过 value 属性值标识组件的名称。例如，cap.service.AdminServiceImpl 通常就需要指定为 adminService。

当在组件类上使用了特定的注解之后，需要在 Spring 的配置文件中作如下的声明：

```
<context: component-scan>;
```

base-package 属性指定一个需要扫描的基类包，Spring 容器将会扫描这个基类包里及其子包中的所有类。下面的示例代码将扫描 cap.dao 子包下的所有类。

```
<context:component-scan base-package="cap.dao.*"></context:component-scan>
```

8.2.2 组件装配

自动装配是指，Spring 在装配 Bean 的时候，根据指定的自动装配规则，将某个 Bean 所需要引用类型的 Bean 注入进来。<context:component-scan> 元素会自动注册 AutowiredAnnotationBeanPostProcessor 实例，可以实现自动装配的注解主要有@Autowired 和@Resource。

1. 使用@Autowired 自动装配 Bean

@Autowired 注解，可以对类成员变量、方法及构造函数进行标注，完成自动装配的工作。通过使用@Autowired 注解来消除 setter，getter 方法。

@Autowired 注解自动装配具有兼容类型（默认采用 byType 自动装配策略）的单个 Bean 属性。

下面将通过一个具体的案例讲解使用@Autowired 自动装配 Bean。

（1）打开 Eclipse 开发环境，选择 Package Explorer，选中工程 spring4 并右击，在弹出的快捷菜单中选择“copy”；然后在 Package Explorer 的空白处右击，在弹出的快捷菜单中选择“paste”；然后在弹出的对话框中的“Project name”输入框中输入工程名：spring5。

（2）修改 src 下的 cap.service.impl 子包中的 AdminServiceImpl.java 类，编辑后的代码如下。

```
package cap.service.impl;
import javax.annotation.Resource;
import org.springframework.beans.factory.annotation.Autowired;
import org.springframework.beans.factory.annotation.Qualifier;
import org.springframework.stereotype.Service;
import cap.bean.Admin;
import cap.service.AdminService;
@Service(value="adminService")
public class AdminServiceImpl implements AdminService {
	@Autowired
	private Admin admin;
	public String sayHello(){
		return "你好：" +admin.getUsername();
	}
}
```

（3）修改 src 中的 cap.test 包，新建 AdminServiceTest，修改后的代码如下。

```
package cap.test;
import org.junit.Test;
import org.springframework.context.ApplicationContext;
import org.springframework.context.support.ClassPathXmlApplicationContext;
import cap.bean.Admin;
import cap.service.AdminService;
public class AdminServiceTest {
private ApplicationContext ctx = new ClassPathXmlApplicationContext("applicationContext.xml");
		@Test
		public void testSayHello(){
			AdminService adminService=(AdminService) ctx.getBean("adminService");
```

```
                System.out.println(adminService.sayHello());
        }
}
```

2. 使用@Resource 自动装配 Bean

@Resource 注解要求提供一个 Bean 名称的属性，若该属性为空，则自动采用标注处的变量或方法名作为 Bean 的名称。

下面将通过一个具体的案例讲解 Spring Bean 自动装配的使用。

（1）在 src 下的 cap.service.impl 子包中新建 AdminServiceImpl 类，编辑后的代码如下。

```
package cap.service.impl;
import javax.annotation.Resource;
import org.springframework.beans.factory.annotation.Autowired;
import org.springframework.beans.factory.annotation.Qualifier;
import org.springframework.stereotype.Service;
import cap.bean.Admin;
import cap.service.AdminService;
@Service(value="adminService")
public class AdminServiceImpl implements AdminService {
        @Resource
        @Qualifier(value="admin")
        private Admin admin;
        public String sayHello(){
                return "你好：" +admin.getUsername();
        }
}
```

代码解释：@Service(value="adminService")此注解相当于在 applicationContext.xml 中定义一个 Bean,id="adminService"，@Resource 和@Qualifier(value="admin")注解指定注入的对象为 admin，使用@Qualifier 注解，自动注入的策略就从 byType 转变成 byName 了。

（2）运行 cap.test 中 AdminServiceTest 测试类的测试方法，实现的效果和前一节相似，请读者自行验证。

第9章 Spring AOP

AOP（Aspect Oriented Programming，面向切面编程），是对传统 OOP（Object Oriented Programming，面向对象编程）的补充。利用 AOP 可以对业务逻辑的各个部分进行隔离，从而使得业务逻辑各部分之间的耦合度降低，提高程序的可重用性，同时提高了开发的效率。

9.1 AOP（面向切面的编程）

9.1.1 AOP 的概念

要充分理解 AOP，首先要了解下面常用的概念，这包括连接点（Joinpoint）、通知（Advice）、切点（Pointcut）、切面（Aspect）、引入（Introduction）、目标对象（Target Object）、AOP 代理（AOP Proxy）和织入（Weaving），下面将分别介绍这些概念。

连接点（Joinpoint）：在程序执行过程中某个特定的点，比如某方法调用的时候或者处理异常的时候。在 Spring AOP 中，一个连接点总是表示一个方法的执行。

通知（Advice）：在切面的某个特定的连接点上执行的动作。其中包括了“around”、“before”和“after”等不同类型的通知。许多 AOP 框架（包括 Spring）都是以拦截器做通知模型，并维护一个以连接点为中心的拦截器链。

切点（Pointcut）：匹配连接点的断言。通知和一个切入点表达式关联，并在满足这个切入点的连接点上运行（例如，当执行某个特定名称的方法时）。切点表达式如何和连接点匹配是 AOP 的核心：Spring 缺省使用 AspectJ 切入点语法。

切面（Aspect）：一个关注点的模块化，这个关注点可能会横切多个对象。事务管理是 Java EE 应用中一个关于横切关注点的很好的例子。在 Spring AOP 中，切面可以使用基于 XML（Schema）或者基于@Aspect 注解的方式来实现。

引入（Introduction）：用来给一个类型声明额外的方法或属性（也被称为连接类型声明（inter-type declaration））。Spring 允许引入新的接口（以及一个对应的实现）到任何被代理的对象。例如，你可以使用引入来使一个 bean 实现 IsModified 接口，以便简化缓存机制。

目标对象（Target Object）：被一个或者多个切面所通知的对象。也被称做被通知（advised）对象。既然 Spring AOP 是通过运行时代理实现的，这个对象永远是一个被代理（proxied）对象。

AOP 代理（AOP Proxy）：AOP 框架创建的对象，用来实现切面契约（例如通知方法执行等等）。在 Spring 中，AOP 代理可以是 JDK 动态代理或者 CGLIB 代理。

织入（Weaving）：把切面连接到其他的应用程序类型或者对象上，并创建一个被通知的对象。根据不同的实现技术，AOP 有三种织入方式：

（1）编译期织入，要求使用特殊的 Java 编译器；

（2）类装载期织入，要求使用特殊的类装载器；

（3）动态代理织入，在运行期为目标类添加增强生成子类的方式。

Spring 采用动态代理织入，而@AspectJ 采用编译期织入和类装载期织入。

9.1.2 AOP 通知类型

通知（Advice）是切入点的可执行代码，AOP 的通知类型主要包括以下 5 种。

（1）前置通知（Before advice）：在某连接点之前执行的通知，但这个通知不能阻止连接点之前的执行流程（除非它抛出一个异常）。

（2）后置通知（After returning advice）：在某连接点正常完成后执行的通知，例如，一个方法没有抛出任何异常，正常返回。

（3）环绕通知（Around advice）：包围一个连接点的通知，如方法调用。这是最强大的一种通知类型。环绕通知可以在方法调用前后完成自定义的行为。它也会选择是否继续执行连接点或直接返回它自己的返回值或抛出异常来结束执行。

（4）异常通知（After throwing advice）：在方法抛出异常退出时执行的通知。

（5）最终通知（After (finally) advice）：当某连接点退出的时候执行的通知（不论是正常返回还是异常退出）。

9.2 Spring AOP 的功能和目标

Spring AOP 使用纯 Java 实现。它不需要专门的编译过程。Spring AOP 不需要控制类装载器层次，因此它适用于 Java EE Web 容器或应用服务器。

Spring 实现 AOP 的方法跟其他的框架不同。Spring 并不是要提供最完整的 AOP 实现，它侧重于提供一种 AOP 实现和 Spring IoC 容器之间的整合，用于帮助解决在企业级开发中的常见问题。

读者可以选择 AspectJ 或者 Spring AOP，以及选择是使用@AspectJ 注解风格还是 Spring XML 配置风格。本章中会分别讲解 Spring AOP 和 AspectJ 的使用。在实际的开发过程中，建议使用 AspectJ 实现 AOP 功能。

9.3 AOP 代理实现

9.3.1 JDK 实现 AOP 代理

Spring 缺省使用 JDK 动态代理（dynamic proxies）来作为 AOP 的代理。这样任何接口（或者接口集）都可以被代理。

下面将通过一个具体的案例讲解使用缺省的 JDK 来作为 AOP 的代理。

（1）打开 Eclipse 开发环境，新建 Java Project，在 Project name 输入框中输入工程名：spring6。

（2）复制工程 spring5 中的 AdminDAO.java 接口，并添加其实现类，编辑后的代码如下。

```
package cap.dao.impl;
import cap.dao.AdminDAO;
public class AdminDAOImpl implements AdminDAO {
        @Override
        public String sayHi() {
                return "欢迎您学习 Spring 技术";
        }
}
```

（3）在 src 的 cap.aop 中新建 JdkProxy 类，编辑后的代码如下。

```
package cap.aop;
import java.lang.reflect.InvocationHandler;
import java.lang.reflect.Method;
import java.lang.reflect.Proxy;
import java.util.Date;
public class JdkProxy implements InvocationHandler{
        /**目标对象**/
        private Object targetObject;
        /**创建代理对象**/
        public Object newProxy(Object targetObject) {
        //将目标对象传入进行代理
                this.targetObject = targetObject;
                return Proxy.newProxyInstance(Thread.currentThread().getContextClassLoader(),
                        targetObject.getClass().getInterfaces(), this);
        }
        @Override
        public Object invoke(Object proxy, Method method, Object[] args) throws Throwable {
                System.out.println("当前日期:"+(new Date()).toString());
                return method.invoke(targetObject, args);
        }
}
```

代码解释：JdkProxy 类实现 InvocationHandler 接口，需要实现 invoke 方法，在 newProxy 方法中通过 Proxy 的静态方法 newProxyInstance 返回一个指定接口的代理类实例，invoke 方法在代理实例上处理方法调用并返回结果。

（4）在 src 中的 cap.test 子包中新建 Junit 测试用例类 AdminDAOTest，编辑后的代码如下。

```
package cap.test;
import org.junit.Test;
import cap.aop.JdkProxy;
import cap.dao.AdminDAO;
import cap.dao.impl.AdminDAOImpl;
public class AdminDAOTest {
	@Test
	public void testSayHi() {
		try {
		JdkProxy proxy = new JdkProxy();
		AdminDAO adminDAO = (AdminDAO) proxy.newProxy(new AdminDAOImpl());
		System.out.println(adminDAO.sayHi());
		} catch (Exception e) {
			e.printStackTrace();
		}
	}
}
```

（5）运行测试用例 testSayHi，在控制台下输出如图 9-1 所示的信息。

```
Problems  @ Javadoc  Declaration  Console
<terminated> AdminDAOTest.testSayHi (4) [JUnit] C:\Program Files\Java\jre7\bin\javaw.exe (2015年6月12日 下午12:06:01)
当前日期:Fri Jun 12 12:06:02 CST 2015
欢迎您学习Spring技术
```

图 9-1 testSayHi 打印信息

9.3.2 CGLIB 实现 AOP 代理

CGLIB（Code Generation Library）是一个强大的、高性能、高质量的 Code 生成类库，它可以在运行期扩展 Java 类与实现 Java 接口。它广泛地被许多 AOP 的框架使用，例如 Spring AOP 和 dynaop，为他们提供方法的 interception（拦截）。CGLIB 包的底层是通过使用一个小而快的字节码处理框架 ASM，来转换字节码并生成新的类。

Spring 也可以使用 CGLIB 代理，对于需要代理类而不是代理接口的时候，CGLIB 代理是很有必要的。如果一个业务对象并没有实现一个接口，默认就会使用 CGLIB。

下面将通过具体的案例讲解 CGLIB 实现 AOP 的代理。

（1）继续在工程 spring6 的 src 中的 cap.aop 新建 CglibProxy 类，编辑后的代码如下。

```
package cap.aop;
import java.lang.reflect.Method;
import java.util.Date;
import net.sf.cglib.proxy.Enhancer;
import net.sf.cglib.proxy.MethodInterceptor;
```

```
import net.sf.cglib.proxy.MethodProxy;
public class CglibProxy implements MethodInterceptor{
    private Object targetObject;// CGLib 需要代理的目标对象
    //创建代理对象
    public Object createProxyObject(Object obj) {
        this.targetObject = obj;
        Enhancer enhancer = new Enhancer();
        enhancer.setSuperclass(obj.getClass());
        enhancer.setCallback(this);
        Object proxyObj = enhancer.create();
        return proxyObj;
    }
    @Override
    public Object intercept(Object proxy, Method method, Object[] args,
            MethodProxy arg3) throws Throwable {
        System.out.println("当前日期:"+(new Date()).toString());
        Object obj = method.invoke(targetObject, args);
        return obj;
    }
}
```

代码解释：本类中实现的功能与使用 JDK 实现代理类似，只是实现的方式不同，本节中采用 CGLIB 实现。

（2）在 src 中的 cap.test 测试类 AdminDAOTest 中添加下面的测试用例。

```
@Test
    public void testCGLIBSayHi() {
        try {
            CglibProxy proxy = new CglibProxy();
    AdminDAO adminDAO = (AdminDAO) proxy.createProxyObject(new AdminDAOImpl());
    System.out.println(adminDAO.sayHi());
        } catch (Exception e) {
            e.printStackTrace();
        }
    }
```

（3）运行测试用例 testCGLIBSayHi，在控制台下输出和上一节类似的信息，请读者自行验证。

9.4 Spring 实现 AOP 代理

Spring 支持 5 种类型的增强，表 9-1 列出了 Spring 支持的 5 种类型增强，这些增强接口都有一些方法，通过实现这些接口方法，就可以将它们织入目标类方法的相应连接点的位置。

表 9-1 Spring 支持的 5 种增强类型

增强类型	描述
前置增强	org.springframework.aop.BeforeAdvice 代表前置增强，因为 Spring 只支持方法级的增强，所以 MethodBeforeAdvice 是目前可用前置增强，表示在目标方法执行前实施增强
后置增强	org.springframework.aop.AfterAdvice 代表后增强，表示目标方法在执行后实施增强
环绕增强	org.springframework.aop.MethodInterceptor 代表环绕增强，表示目标方法执行前后实施增强
异常抛出增强	org.springframework.aop.ThrowsAdvice 代表抛出异常增强，表示目标方法抛出异常后实施增强
引介增强	org.springframework.aop.IntroductionInterceptor 代表引介增强，表示在目标类中添加一些新的方法和属性

9.4.1 ProxyFactoryBean 实现 AOP 代理

ProxyFactoryBean 负责为其他 Bean 创建代理实例，其内部使用 ProxyFactory 来完成工作，下面是 ProxyFactoryBean 常用的属性。

- target：代理的目标对象
- proxyInterfaces：代理所要实现的接口，可以是多个接口。
- interceptorNames：需要植入目标对象的 Bean 列表。
- singleton：返回的代理是否是单实例，默认为单实例。
- optimize：是否对创建的代理进行优化，当设置属性为 true 时，强制使用 CGLIB。
- ProxyTargetClass：是否代理目标类，而不是对接口进行代理。设置为 true 时候，使用 CGLIB 代理。

下面通过一个具体的例子讲解使用 ProxyFactoryBean 作为实现 AOP 代理。

（1）打开 Eclipse 开发环境，选择 Package Explorer，选中工程 spring5 并右击，在弹出的快捷菜单中选择“copy”，然后在 Package Explorer 的空白处右击，在弹出的菜单中选择“paste”，然后在弹出的对话框中的 Project name 输入框处输入工程名：Spring6x0。

（2）在工程 spring6 中 src 的 cap.util 新建 AdminInterceptor 类，编辑后的代码如下。

```
package cap.util;
import java.lang.reflect.Method;
import java.util.Date;
import org.aopalliance.intercept.MethodInterceptor;
import org.aopalliance.intercept.MethodInvocation;
import org.springframework.aop.AfterReturningAdvice;
import org.springframework.aop.MethodBeforeAdvice;
public class AdminInterceptor implements MethodBeforeAdvice,AfterReturningAdvice,MethodInterceptor{
    @Override
    public void afterReturning(Object arg0, Method arg1, Object[] arg2,
                Object arg3) throws Throwable {
        System.out.println("记录前:现在时间是:"+new Date().toString());
    }
    @Override
    public void before(Method arg0, Object[] arg1, Object arg2)
                throws Throwable {
```

```
            System.out.println("记录前:现在时间是:"+new Date().toString());
      }
      @Override
      public Object invoke(MethodInvocation mi) throws Throwable {
            String info = mi.getMethod().getDeclaringClass() + "."
                        + mi.getMethod().getName() + "()";
            System.out.println(info);
            try {
                  Object result = mi.proceed();
                  return result;
            } finally {
                  System.out.println(info);
            }
      }
}
```

代码解释：afterReturning 方法定义后置增强，before 方法定义前置增强，invoke 方法定义环绕增强。

（3）修改 src 下的配置文件 applicationContext.xml，添加下面的代码。

```
<?xml version="1.0" encoding="UTF-8"?>
<beans xmlns="http://www.springframework.org/schema/beans"
    xmlns:xsi="http://www.w3.org/2001/XMLSchema-instance"
    xmlns:context="http://www.springframework.org/schema/context"
    xmlns:aop="http://www.springframework.org/schema/aop"
    xmlns:tx="http://www.springframework.org/schema/tx"
    xsi:schemaLocation="
     http://www.springframework.org/schema/beans
     http://www.springframework.org/schema/beans/spring-beans.xsd
     http://www.springframework.org/schema/tx
     http://www.springframework.org/schema/tx/spring-tx.xsd
    http://www.springframework.org/schema/context
     http://www.springframework.org/schema/context/spring-context.xsd
     http://www.springframework.org/schema/aop
     http://www.springframework.org/schema/aop/spring-aop.xsd">
  <bean id="admin" class="cap.bean.Admin">
    <property name="id" value="1"/>
    <property name="username" value="starlee2008"/>
    <property name="password" value="starlee2008"/>
  </bean>
  <bean id="adminDAO" class="cap.dao.impl.AdminDAOImpl"><!—代码①   -->
  <property name="admin" ref="admin"></property>
  </bean>
  <!—代码②-->
  <bean id="adminInterceptor" class="cap.util.AdminInterceptor"></bean>

  <bean id="logProxy" class="org.springframework.aop.framework.ProxyFactoryBean">
        <property name="proxyInterfaces"> <!—代码③   -->
```

```
                <value>cap.dao.AdminDAO</value>
            </property>
            <property name="target"><!—代码④  -->
            <ref bean="adminDAO"/>
            </property>
             <property name="interceptorNames">  <!—代码⑤  -->
                <list>
                    <value>adminInterceptor</value>
                </list>
            </property>
        </bean>
    </beans>
```

代码解释：代码①定义需要代理的 Bean 对象 adminDAO，代码②定义增强类，代码③指定代理的接口，如果是多个接口，使用<list>元素，代码④指定需要代理的 Bean 对象，代码⑤指定需要植入目标对象的 Bean 列表。

运行测试方法，继续运行 testSayHi 方法，在控制台会出现如图 9-2 所示类似的信息。

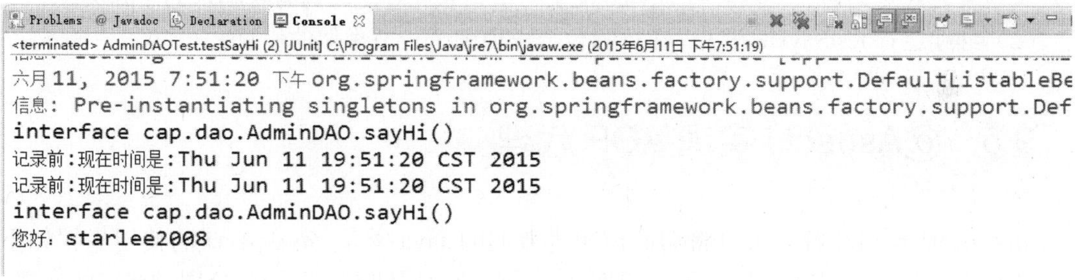

```
Problems  @ Javadoc  Declaration  Console
<terminated> AdminDAOTest.testSayHi (2) [JUnit] C:\Program Files\Java\jre7\bin\javaw.exe (2015年6月11日 下午7:51:19)
六月 11, 2015 7:51:20 下午 org.springframework.beans.factory.support.DefaultListableBe
信息: Pre-instantiating singletons in org.springframework.beans.factory.support.Def
interface cap.dao.AdminDAO.sayHi()
记录前:现在时间是:Thu Jun 11 19:51:20 CST 2015
记录前:现在时间是:Thu Jun 11 19:51:20 CST 2015
interface cap.dao.AdminDAO.sayHi()
您好: starlee2008
```

图 9-2　testSayHi 打印信息

9.4.2　AOP 自动代理

通过 ProxyFactoryBean 创建织入切面的代理，每个需要被代理的 Bean 都需要使用一个 ProxyFactoryBean 进行配置，如果有很多需要代理的 Bean，配置就很麻烦，Spring 提供了自动代理机制，让容器自动生成代理，把用户从繁琐的配置中解脱出来，在内部，Spring 使用 BeanPostProcessor 自动完成工作。

在本节中主要使用 DefaultAdvisorAutoProxyCreator 实现目标 Bean 自动创建代理，DefaultAdvisorAutoProxyCreator 能够扫描容器中的 Advisor，并将 Advisor 自动织入匹配的目标 Bean 中，即为匹配的目标 Bean 自动创建代理。

下面将通过一个具体的实例讲解使用 DefaultAdvisorAutoProxyCreator 为目标 Bean 自动创建代理。

（1）修改 src 下的配置文件 applicationContext.xml，修改后的代码如下。

```
<bean id="adminDAO" class="cap.dao.impl.AdminDAOImpl">
    <property name="admin" ref="admin"></property>
</bean>
    <!—代码①  -->
```

```
<bean id="autoProxyCreator" class="org.springframework.aop.framework.autoproxy.
DefaultAdvisorAutoProxyCreator">
</bean>
<!—代码②  -->
<bean id="advisor" class="org.springframework.aop.support.RegexpMethodPointcutAdvisor">
  <property name="pattern">
    <value>.*DAOImpl.say*.*</value>   <!-- 业务实现方法名匹配 -->
  </property>
  <property name="advice">
    <ref bean="adminInterceptor"/>
  </property>
</bean>
<!—代码③  -->
<bean id="adminInterceptor" class="cap.util.AdminInterceptor"></bean>
```

代码解释：代码①使用 DefaultAdvisorAutoProxyCreator 创建一个 Bean，它负责将容器中的 Advisor 织入匹配的目标 Bean 中。代码②使用 RegexpMethodPointcutAdvisor 定义一个切点，使用正则表达式定义切点，然后将 advice 设置为引用代码③定义的需要插入的增强类。

（2）继续运行 testSayHi 方法。在控制台会出现和上一节类似的信息。

9.5 @AspectJ 实现 AOP 代理

@AspectJ 使用注解，可以将切面声明为普通的 Java 类。 @AspectJ 的使用步骤如下：①需要启用@AspectJ 支持；②需要声明切面；③需要声明切点。下面将分别讲述这些步骤。

9.5.1 启用@AspectJ

在 Spring 配置中使用@AspectJ 切面，必须启用 Spring 对@AspectJ 切面配置的支持，并确保自动代理（autoproxying）的 Bean 是否能被这些切面通知。自动代理是指 Spring 会判断一个 Bean 是否使用了一个或多个切面通知，并据此自动生成相应的代理以拦截其方法调用，并且确保通知在需要时执行。

首先在工程中的 applicationContext.xml 添加下面的代码：启用 Spring 对@AspectJ 的支持。

```
<aop:aspectj-autoproxy/>
```

同时需要在工程的 classpath 中引入两个 AspectJ 库：aspectjweaver.jar 和 aspectjrt.jar。这些库可以在 AspectJ 的安装包的 lib 目录里找到。

9.5.2 声明切面（Aspect）

启用@AspectJ 支持后，在 Spring 的配置文件中定义的任意带有一个@Aspect 切面（拥有@Aspect 注解）的 Bean 都将被 Spring 自动识别并用于配置 Spring AOP。下面的例子展示了切面的定义。

applicationContext.xml 中一个常见的 Bean 定义，它指向一个使用了@Aspect 注解的 Bean 类。

```
<bean id="aspectj" class="cap.util.AdminAspectjLogger" />
```

除了使用@Aspect 注解，Bean 定义还使用 AdminAspectjLogger 类的定义，使用 org.aspectj.lang.annotation.Aspect 注解。

```
package cap.util;
import org.aspectj.lang.annotation.Aspect;
@Aspect
public class AdminAspectjLogger {
}
```

9.5.3 声明切点（pointcut）

一个切点声明有两个部分：一个包含名字和任意参数的签名，还有一个切入点表达式。该表达式决定了我们关注那个方法的执行。在@AspectJ 注解风格的 AOP 中，一个切入点签名通过一个普通的方法定义来提供，并且切入点表达式使用@Pointcut 注解来表示（作为切入点签名的方法必须返回 void 类型）。表 9-2 显示了常用@AspectJ 切点表达式函数的说明。

表 9-2 切点函数

函 数	描 述
execution	使用“execution(方法表达式)”匹配方法执行
@annotation	使用“@annotation(注解类型)”匹配当前执行方法持有指定注解的方法；注解类型也必须是全限定类型名
within	使用“within(类型表达式)”匹配指定类型内的方法执行
this	使用“this(类型全限定名)”匹配当前 AOP 代理对象类型的执行方法
target	使用“target(类型全限定名)”匹配当前目标对象类型的执行方法
args	使用“args(参数类型列表)”匹配当前执行的方法传入的参数为指定类型的执行方法

有些@AspectJ 切点表达式中的方法参数支持通配符：主要有*、..和+，下面将讲述这些通配符的含义。

*：匹配任何数量字符。

..：匹配任何数量字符的重复，如在类型模式中匹配任何数量子包；而在方法参数模式中匹配任何数量参数。

+：匹配指定类型的子类型；仅能作为后缀放在类型模式后边。

9.5.4 @AspectJ 实现 AOP 代理实例

同样基于@AspectJ 的 AOP 也分为基于 XML 和基于注解方式实现，下面将分别介绍这两种方式的实现。

1. XML 方式实现 AOP

下面将通过一个实例讲解基于 XML(Schema)实现 AOP 代理，实现的步骤如下。

（1）打开 Eclipse 开发环境，选择 Package Explorer，选中工程 spring5 并右击，在弹出的快捷菜单中选择“copy”，然后在 Package Explorer 的空白处右击，在弹出的菜单中选择“paste”，然后在弹出的对话框中的 Project name 输入框处输入工程名：spring6x1。

（2）在工程 Spring6x1 中 src 的 cap.util 新建 AdminLogger 类，编辑后的代码如下。

```
package cap.uitl;
import java.util.Date;
import org.aspectj.lang.ProceedingJoinPoint;
public class AdminLogger {
    public void logBefore() {
        System.out.println("记录前:现在时间是:"+new Date().toString());
    }
    // spring 中 After 通知
    public void logAfter() {
        System.out.println("记录前:现在时间是:"+new Date().toString());
    }
    // spring 中 Around 通知
    public Object logAround(ProceedingJoinPoint joinPoint) {
        // 方法执行前的代理处理
        System.out.println("记录开始:现在时间是:"+new Date().toString());
        Object[] args = joinPoint.getArgs();
        Object obj = null;
        try {
            obj = joinPoint.proceed(args);
        } catch (Throwable e) {
            e.printStackTrace();
        }
        // 方法执行后的代理处理
        System.out.println("记录结束结束:现在时间是:"+new Date().toString());
        return obj;
    }
  }
}
```

代码解释：logBefore 方法定义了前置增强，logAfter 方法定义后置增强，logAround 方法定义了环绕增强。

（3）修改 src 下的配置文件 applicationContext.xml，添加下面的代码。

```
<bean id="advice" class="cap.aop.AdminLogger" /><!—代码①  -->
    <aop:config>
        <aop:aspect ref="advice"><!—代码②  -->
            <aop:pointcut id="pointcut" expression="execution(* cap.dao.impl.*.*(..))" /><!—代码③  -->
            <aop:before method="logBefore" pointcut-ref="pointcut" /><!—代码④  -->
            <aop:after method="logAfter" pointcut-ref="pointcut" /><!—代码⑤  -->
            <aop:around method="logAround" pointcut-ref="pointcut" /><!—代码⑥  -->
        </aop:aspect>
    </aop:config>
```

代码解释：代码①定义了增强类 advice 对象，代码②定义 AOP 的切面，需要引用代码①定义的 advice 对象，代码③定义切点，其中切点的表达式使用 execution，其参数使用前面介绍的通配符。其中第一个*代表任意的返回类型，第二个*代表 cap.dao.impl 子包下面的全部类，第三个*代表类中的任意方法，..表示 cap.dao.impl 子包中类的任何方法，不管参数如何（匹配任意的参数）。代码④指定前置通知执行的方法，需要引用代码③定义的切点，代码⑤指定后置通知执行的方法，需要引用代码③定义的切点，代码⑥指定环绕通知执行的方法，需要引用代码③定义的切点。

（4）运行测试方法，继续运行 testSayHi 方法。在控制台会出现如图 9-3 所示类似的信息。

```
Markers  Properties  Servers  Data Source Explorer  Snippets  Console
<terminated> AdminDAOTest.testSayHi (1) [JUnit] C:\Program Files\Java\jre7\bin\javaw.exe (2015年5月17日 上午10:08:55
记录结束结束:现在时间是:Sun May 17 10:08:58 CST 2015
记录前:现在时间是:Sun May 17 10:08:58 CST 2015
您好：starlee2008
```

图 9-3 testSayHi 打印信息

2. 注解方式实现 AOP

@AspectJ 为各种增强类型提供了不同的注解，它们位于 org.aspectj.lang.annotation.*包中，这些注解类拥有若干成员，可以通过这些成员完成定义切点信息，绑定连接点参数等操作，@AspectJ 所提供的常用增强注解如表 9-3 所示。

表 9-3 @AspectJ 提供的常用增强注解

注 解	描 述
@Before	表示前置增强，相当于 BeforeAdvice 的功能
@After	表示最终增强，不管是抛出异常还是正常退出，该增强都会执行
@AfterThrowing	表示抛出增强，相当于 ThrowsAdvice
@AfterRunning	表示后置增强，相当于 AfterReturningAdvice
@Around	表示环绕增强，相当于 MethodInterceptor

上面一节讲解了 AspectJ 基于 XML 实现 AOP 代理的使用，在本节中继续讲解基于注解实现 AOP 代理的使用。

（1）在工程 Spring6x1 的 cap.util 包下新建 AdminAspectjLogger 类，编辑后的代码如下。

```
package cdap.aop;
import java.util.*;
import org.aspectj.lang.ProceedingJoinPoint;
import org.aspectj.lang.annotation.After;
import org.aspectj.lang.annotation.Around;
import org.aspectj.lang.annotation.Aspect;
import org.aspectj.lang.annotation.Before;
@Aspect
public class AdminAspectjLogger {
    /**
     * 必须为 final String 类型的,注解里要使用的变量只能是静态常量类型的
     */
    public static final String AAP= "execution(* cap.dao.impl.*.*(..))";
    @Before(AAP)
    public void logBefore() {
        System.out.println("记录前:现在时间是:"+new Date().toString());
```

```
    }
    @After(AAP)
    public void logAfter() {
        System.out.println("记录前:现在时间是:"+new Date().toString());
    }
    @Around(AAP)    //spring 中 Around 通知
    public Object logAround(ProceedingJoinPoint joinPoint) {
        System.out.println("记录开始:现在时间是:"+new Date().toString()); // 方法执行前的代理处理
        Object[] args = joinPoint.getArgs();
        Object obj = null;
        try {
            obj = joinPoint.proceed(args);
        } catch (Throwable e) {
            e.printStackTrace();
        }
        System.out.println("记录结束结束:现在时间是:"+new Date().toString());
// 方法执行后的代理处理
        return obj;
    }
}
```

代码解释：注解@Before(AAP)定义了前置增强，@After(AAP)定义最终增强，@Around(AAP)定义环绕增强。

（2）修改 src 中的配置文件 applicationContext.xml，对 aop 的标签注释，并添加下面的代码。

```
<aop:aspectj-autoproxy/>
<bean id="aspectj" class="cap.util.AdminAspectjLogger" />
```

（3）运行测试类中的 testSayHi 方法，运行的结果和前一节相似，只是显示的时间不同。

第10章 Spring JDBC

10.1 Spring JDBC

Spring JDBC 是 Spring 提供的持久层技术。该技术的出现可以降低使用 JDBC API 的标准，从而以一种更直接、更简洁的方式使用 JDBC API。

回顾 Struts2 应用中使用 Struts2+JDBC 实现“增、删、改、查”的例子，我们需要做以下的工作：

- 指定数据库连接参数；
- 打开数据库连接；
- 声明 SQL 语句；
- 预编译并执行 SQL 语句；
- 遍历查询结果（如果需要的话）；
- 处理每一次遍历操作；
- 处理抛出的任何异常；
- 处理事务；
- 关闭数据库连接。

而使用 Spring JDBC 之后，上面的步骤就可以简化到只需要实现声明 SQL 语句和处理每一次遍历操作。Spring 将替我们完成所有的 JDBC 底层细节处理工作。

10.2 Spring JDBC 包结构

Spring JDBC 抽象框架由四个包构成：core、dataSource、object 以及 support。

org.springframework.jdbc.core 包由 JdbcTemplate 类以及相关的回调接口（callback interface）

和类组成。

org.springframework.jdbc.datasource 包由一些用来简化 DataSource 访问的工具类，以及各种 DataSource 接口的简单实现（主要用于单元测试以及在 Java EE 容器之外使用 JDBC）组成。工具类提供了一些静态方法，诸如通过 JNDI 获取数据连接以及在必要的情况下关闭这些连接。它支持绑定线程的连接，比如被用于 DataSourceTransactionManager 的连接。

org.springframework.jdbc.object 包由封装了查询、更新以及存储过程的类组成，这些类的对象都是线程安全并且可重复使用的。

org.springframework.jdbc.support 包提供了一些 SQLException 的转换类以及相关的工具类。

10.3 DataSource 接口

为了从数据库中取得数据，首先需要获取一个数据库连接。Spring 通过 DataSource 对象来完成这个工作。 DataSource 是 JDBC 规范的一部分， 它被视为一个通用的数据库连接工厂。通过使用 DataSource，Container 或 Framework 可以将连接池以及事务管理的细节从应用代码中分离出来。

在使用 Spring JDBC 时，可以通过 JNDI 获得数据源，也可以自行配置数据源（使用 Spring 提供的 DataSource 实现类）。使用后者可以更方便地脱离 Web 容器来进行单元测试。

下面我们将通过一个例子来说明如何配置 DriverManagerDataSource。

（1）在 Eclipse 中新建 Java Project，工程名为 spring7，在 src 的 cap.util 下新建 DBUtil.java 类，编辑后的代码如下。

```
package cdap.util;
import org.springframework.jdbc.datasource.DriverManagerDataSource;
public class DBUtil {
	static DriverManagerDataSource dataSource=null;
	public static DriverManagerDataSource getDS()
	{
		dataSource = new DriverManagerDataSource();
		dataSource.setDriverClassName("com.mysql.jdbc.Driver");
		dataSource.setUrl("jdbc:mysql://localhost:3306/cap");
		dataSource.setUsername("root");
		dataSource.setPassword("admin");
		return dataSource;
	}
}
```

代码解释：静态方法 getDS 中，首先实例化 DriverManagerDataSource 的对象 dataSource，然后通过 setter 方法设置数据源的属性，包括 JDBC 驱动、连接 URL、用户名和密码。

（2）在 src 的 cap .dao 中新建 AdminDAO 接口。

```
package cap.dao;
import java.util.List;
import cap.bean.Admin;
public interface AdminDAO {
```

```
    public List<Admin> findAdmins();
}
```

（2）在 src 的 cap.dao.impl 中新建 AdminDAO 接口的实现类 AdminDAOImpl.java。

```
package cap.dao.impl;
import java.util.List;
import org.springframework.jdbc.core.JdbcTemplate;
import org.springframework.jdbc.core.simple.ParameterizedBeanPropertyRowMapper;
import cap.bean.Admin;
import cap.dao.AdminDAO;
import cap.util.DBUtil;
public class AdminDAOImpl implements AdminDAO {
    JdbcTemplate jdbcTemplate=new JdbcTemplate(DBUtil.getDS());
    @Override
    public List<Admin> findAdmins() {
        String sql="select * from admin order by id";
        return jdbcTemplate.query(sql, ParameterizedBeanPropertyRowMapper.newInstance(Admin.class));
        }
}
```

（4）在 src 的 cap.test 包中新建 AdminDAOTest 测试类。

```
package cap.test;
import java.util.List;
import org.junit.Test;
import org.springframework.context.ApplicationContext;
import org.springframework.context.support.ClassPathXmlApplicationContext;
import cap.bean.Admin;
import cap.dao.AdminDAO;
public class AdminDAOTest {
    private ApplicationContext ctx = new ClassPathXmlApplicationContext("applicationContext.xml");
    @Test
    public void testFindAdmins() {
        AdminDAO adminDAO=(AdminDAO) ctx.getBean("adminDAO");
        List<Admin> adminList=adminDAO.findAdmins();
        for(Admin admin:adminList){
            System.out.println(admin.getUsername());
        }
    }
}
```

10.4　JdbcTemplate 类

JdbcTemplate 是 core 包的核心类。它完成了资源的创建以及释放工作，从而简化了我们对 JDBC 的使用，还可以帮助我们避免一些常见的错误，比如忘记关闭数据库连接。

JdbcTemplate 将完成 JDBC 核心处理流程，比如 SQL 语句的创建、执行，而把 SQL 语句的生成以及查询结果的提取工作留给我们的应用代码。它可以完成 SQL 查询、更新以及调用存储过程，可以对 ResultSet 进行遍历并加以提取。它还可以捕获 JDBC 异常并将其转换成 org.springframework.dao 包中定义的通用的、信息更丰富的异常。

JdbcTemplate 主要提供以下五类方法。

- execute 方法：可以用于执行任何 SQL 语句，一般用于执行 DDL 语句；
- update 方法：用于执行新增、修改、删除等语句；
- batchUpdate 方法：用于执行批处理相关语句；
- query 方法及 queryForXXX 方法：用于执行查询相关语句：queryForList、queryForMap、queryForObject、queryForRowSet;
- call 方法：用于执行存储过程、函数相关语句。

上述方法的定义可以参考 Spring Document，由于篇幅的原因不在此一一列举。

10.4.1 使用 JdbcTemplate

通过在 DAO 类中使用 JdbcTemplate 的两种方式：DAO 类中定义 JdbcTemplate 和 DAO 类继承 JdbcDaoSupport。

10.4.2 DAO 类中定义 JdbcTemplate

DAO 类中定义 JdbcTemplate 的对象 jdbcTemplate，然后注入 JdbcTemplate 的对象。使用的案例如下。

（1）打开 Eclipse 开发环境，选择 Package Explorer，选中工程 spring7 并右击，在弹出的快捷菜单中选择“copy”，然后在 Package Explorer 的空白处右击，在弹出的菜单中选择“paste”，然后在弹出的对话框中的 Project name 输入框处输入工程名：spring8x0。

（2）修改 src 中 cap.dao.impl 下的 AdminDAOImpl 类。

```
package cap.dao.impl;
import java.util.List;
import org.springframework.beans.factory.annotation.Autowired;
import org.springframework.jdbc.core.BeanPropertyRowMapper;
import org.springframework.jdbc.core.JdbcTemplate;
import org.springframework.jdbc.core.RowMapper;
import org.springframework.stereotype.Repository;
import cap.bean.Admin;
import cap.dao.AdminDao;
@Repository("adminDAO")
public class AdminDaoJdbcTemplateImpl implements AdminDao {
    @Autowired
    private JdbcTemplate jdbcTemplate;
    public JdbcTemplate getJdbcTemplate() {
        return jdbcTemplate;
    }

    public void setJdbcTemplate(JdbcTemplate jdbcTemplate) {
```

```
        this.jdbcTemplate = jdbcTemplate;
    }
    @Override
    public int addAdmin(Admin admin) {
        String sql = "insert into admin(user_name,password) values(?,?)";
        return jdbcTemplate.update(sql, admin.getUsername(),admin.getPassword());
    }
    @Override
    public int updateAdmin(Admin admin) {
        String sql = "update admin set user_name=?,password=? where id=?";
        return jdbcTemplate.update(sql, admin.getUsername(),admin.getPassword(), admin.getId());
    }
    @Override
    public int deleteAdmin(Integer id) {
        String sql = "delete from admin where id=?";
        return jdbcTemplate.update(sql, id);
    }
    @Override
    public Admin findById(int id) {
        String sql = "SELECT id,user_name AS username,password FROM admin WHERE id = ?";
        RowMapper<Admin> rowMapper = new BeanPropertyRowMapper<>(Admin.class);
        Admin admin = jdbcTemplate.queryForObject(sql, rowMapper, id);
        return admin;
    }

    @Override
    public List<Admin> findAdmins() {
        String sql = "SELECT id,user_name AS username,password FROM admin ORDER BY id";
        RowMapper<Admin> rowMapper = new BeanPropertyRowMapper<>(Admin.class);
        List<Admin> admins = jdbcTemplate.query(sql, rowMapper);
        return admins;
    }
}
```

（3）修改 src 下的 Spring 配置文件 applicationContext.xml，编辑后的代码如下。

```
<?xml version="1.0" encoding="UTF-8"?>
<beans xmlns="http://www.springframework.org/schema/beans"
    xmlns:xsi="http://www.w3.org/2001/XMLSchema-instance" xmlns:aop="http://www.springframework.
                                                                org/schema/aop"
    xmlns:context="http://www.springframework.org/schema/context"
    xsi:schemaLocation="http://www.springframework.org/schema/beans http://www.springframework.org/
                                                                schema/beans/spring-beans.xsd
        http://www.springframework.org/schema/aop http://www.springframework.org/schema/aop/spring-
                                                                aop.xsd
        http://www.springframework.org/schema/context http://www.springframework.org/schema/context/
                                                                spring-context.xsd">
    <!-- 自动扫描包 -->
    <context:component-scan base-package="cap.dao"></context:component-scan>
```

```
    <!-- 配置数据源 -->
    <bean id="dataSource"
        class="org.springframework.jdbc.datasource.DriverManagerDataSource">
        <property name="driverClassName" value="com.mysql.jdbc.Driver" />
        <property name="url" value="jdbc:mysql://localhost:3306/cap" />
        <property name="username" value="root" />
        <property name="password" value="root" />
    </bean>
    <!-- 配置 Spring 的 JdbcTemplate -->
    <bean id="jdbcTemplate" class="org.springframework.jdbc.core.JdbcTemplate">
        <property name="dataSource" ref="dataSource"></property>
    </bean>
</beans>
```

（4）运行 cap.test 中 AdminDAOTest 测试类中的 testFindAdmins 方法，运行的结果和前一节相似，在此不再说明。

10.4.3 DAO 类继承 JdbcDaoSupport

在 DAO 类中使用 JdbcTemplate 的第二种方法是在 DAO 中继承 JdbcDaoSupport，然后注入 DataSource。下面将通过具体的案例进行讲解。

（1）打开 Eclipse 开发环境，选择 Package Explorer，选中工程 spring8x0 并右击，在弹出的快捷菜单中选择“copy”，然后在 Package Explorer 的空白处右击，在弹出的菜单中选择“paste”，然后在弹出的对话框中的 Project name 输入框处输入工程名：spring8x1。

（2）修改 src 中 cap.dao.impl 的 AdminDAOImpl 类，修改后的代码如下。

```
package cap.dao.impl;
import java.util.List;
import org.springframework.jdbc.core.simple.ParameterizedBeanPropertyRowMapper;
import org.springframework.jdbc.core.support.JdbcDaoSupport;
import cap.bean.Admin;
import cap.dao.AdminDAO;
public class AdminDAOImpl extends JdbcDaoSupport implements AdminDAO {
    @Override
    public List<Admin> findAdmins() {
        String sql="select * from admin order by id";
            return this.getJdbcTemplate().query(sql,
ParameterizedBeanPropertyRowMapper.newInstance(Admin.class));
        }
    //省略部分方法
}
```

（3）修改 src 的 applicationContext.xml 中的内容，修改后的配置文件如下。

```
<?xml version="1.0" encoding="UTF-8"?>
<beans xmlns="http://www.springframework.org/schema/beans"
    xmlns:xsi="http://www.w3.org/2001/XMLSchema-instance"
    xmlns:context="http://www.springframework.org/schema/context"
```

```
        xmlns:aop="http://www.springframework.org/schema/aop"
        xmlns:tx="http://www.springframework.org/schema/tx"
        xsi:schemaLocation="
         http://www.springframework.org/schema/beans
         http://www.springframework.org/schema/beans/spring-beans.xsd
         http://www.springframework.org/schema/tx
         http://www.springframework.org/schema/tx/spring-tx.xsd
       http://www.springframework.org/schema/context
        http://www.springframework.org/schema/context/spring-context.xsd
         http://www.springframework.org/schema/aop
         http://www.springframework.org/schema/aop/spring-aop.xsd">
        <!-- 配置数据源 -->
        <bean id="dataSource" class="org.springframework.jdbc.datasource.DriverManagerDataSource">
              <property name="driverClassName" value="com.mysql.jdbc.Driver"/>
              <property name="url" value="jdbc:mysql://localhost:3306/cap"/>
              <property name="username" value="root"/>
              <property name="password" value="admin"/>
        </bean>
    <!—在 adminDAO 中注入数据源 -->
        <bean id="adminDAO" class="cap.dao.impl.AdminDAOImpl">
        <property name="dataSource" ref="dataSource"></property>
        </bean>
    </beans>
```

（4）运行 cap.test 中 AdminDAOTest 测试类中的 testFindAdmins 方法，运行的结果和前一节相似。

通过对 JdbcTemplate 两种使用方法的讲解，可以看到采用继承 JdbcDaoSupport 的第二种方法比第一种方法稍简单，在实际开发过程中，读者可以根据实际情况选择其中的一种方法实现。

10.5 NamedParameterJdbcTemplate 类

在传统的 JDBC 用法中，SQL 参数都是用“?”占位符表示的，并且受到位置的限制。定位参数的问题在于，一旦参数的顺序发生变化，就必须改变参数绑定。

在 Spring JDBC 框架中，绑定 SQL 参数的另一种选择是使用命名参数(named parameter)。

命名参数：SQL 按名称(以冒号开头)而不是按位置进行指定。命名参数更易于维护，也提升了可读性。命名参数由框架类在运行时用占位符取代。

Spring JDBC 框架中的 NamedParameterJdbcTemplate 类增加了在 SQL 语句中使用命名参数的支持。NamedParameterJdbcTemplate 类内部封装了一个普通的 JdbcTemplate，并作为其代理来完成大部分工作。

NamedParameterJdbcTemplate 类是线程安全的，该类的最佳使用方式不是每次操作的时候实例化一个新的 NamedParameterJdbcTemplate，而是针对每个 DataSource 只配置一个 NamedParameterJdbcTemplate 实例（比如在 Spring IoC 容器中使用 Spring IoC 来进行配置），然后在那些使用该类的 DAO 中共享该实例。

在 Spring IoC 容器中配置 NamedParameterJdbcTemplate 时，由于其没有无参构造器，所以必须为其构造器指定参数，配置代码如下。

```
<bean id="namedParameterJdbcTemplate" class="org.springframework.jdbc.core.namedparam.
                                    NamedParameterJdbcTemplate">
                        <constructor-arg ref="dataSource"></constructor-arg>
</bean>
```

在 SQL 语句中使用命名参数时，可以在一个 Map 中提供参数值，参数名为键。

例如使用 **int** update(String sql, Map<String, ?> paramMap)方法进行更新操作。

```
public int addAdmin(Admin admin) {
        String sql = "insert into admin(user_name,password) values(:username,:password)";
            Map<String, Object> params = new HashMap<String, Object>();
           params.put("username", admin.getUsername());
            params.put("password", admin.getPassword());
            return namedParameterJdbcTemplate.update(sql, params);
}
```

在上面例子中，sql 变量使用了命名参数占位符 username 和 password，使用基于 Map 风格的名值对将命名参数传递给 NamedParameterJdbcTemplate。使用这种方式时，如有多个参数，则不用去对应位置，而是直接对应参数名，这样便于维护，缺点是使用比较麻烦。

使用命名参数时，还可以使用 SqlParameterSource 提供参数值。可以使用 SqlParameterSource 的 BeanPropertySqlParameterSource 实现类作为参数，要求 SQL 语句中的参数名和类的属性一致。

例如使用 **int** update(String sql, SqlParameterSource paramSource)方法进行更新操作。

```
public int addAdmin(Admin admin) {
        String sql = "insert into admin(username,password) values(:username,:password)";
           SqlParameterSource parameterSource = new BeanPropertySqlParameterSource(admin);
            return namedParameterJdbcTemplate.update(sql, parameterSource);
}
```

下面将通过一个具体的实例讲解使用 NamedParameterJdbcTemplate 实现“增、删、改、查”操作。

（1）打开 Eclipse 开发环境，选择 Package Explorer，选中工程 spring7 并右击，在弹出的快捷菜单中选择“copy”，然后在 Package Explorer 的空白处右击，在弹出的菜单里选择“paste”，然后在弹出的对话框中的 Project name 输入框处输入工程名：spring9。

（2）修改 src 下 cap.dao.impl 中的 AdminDaoNamedParameterJdbcTemplateImpl.java 类。

```
package cap.dao.impl;
import java.util.HashMap;
import java.util.List;
import java.util.Map;
import javax.sql.DataSource;
import org.springframework.jdbc.core.namedparam.BeanPropertySqlParameterSource;
import org.springframework.jdbc.core.namedparam.NamedParameterJdbcTemplate;
import org.springframework.jdbc.core.namedparam.SqlParameterSource;
```

```
import org.springframework.jdbc.core.simple.ParameterizedBeanPropertyRowMapper;
import cap.bean.Admin;
import cap.dao.AdminDAO;
public class AdminDAOImpl implements AdminDAO {
    private NamedParameterJdbcTemplate namedJdbcTemplate;
    public void setDataSource(DataSource dataSource) {
        this.namedJdbcTemplate = new NamedParameterJdbcTemplate(dataSource);
    }
    @Override
        public int addAdmin(Admin admin) {
            String sql = "insert into admin(username,password) values(:username,:password)";
            /*
            Map<String, Object> params = new HashMap<String, Object>();
            params.put("username", admin.getUsername());
            params.put("password", admin.getPassword());
            */
            SqlParameterSource parameterSource = new BeanPropertySqlParameterSource(admin);
            return namedJdbcTemplate.update(sql, parameterSource);
        }

        @Override
        public int updateAdmin(Admin admin) {
            String sql = "update admin set username=:username,password=:password where id=:id";
            Map<String, Object> params = new HashMap<String, Object>();
            params.put("username", admin.getUsername());
            params.put("password", admin.getPassword());
            params.put("id", admin.getId());
            return namedJdbcTemplate.update(sql, params);
        }
        @Override
        public int deleteAdmin(Integer id) {
            Map<String, Object> params = new HashMap<String, Object>();
            params.put("id", id);
            String sql = "delete from admin where id=:id";
            return namedJdbcTemplate.update(sql, params);
        }
        @Override
        public Admin findById(Integer id) {
            String sql = "select * from admin where id=:id";
            Map<String, Object> params = new HashMap<String, Object>();
            params.put("id", id);
            return namedJdbcTemplate.queryForObject(sql, params, ParameterizedBeanPropertyRowMapper.
            newInstance(Admin.class));
        }
        @Override
        public List<Admin> findAdmins() {
            String sql = "select * from admin order by id";
            return namedJdbcTemplate.query(sql,
```

```
                        ParameterizedBeanPropertyRowMapper.newInstance(Admin.class));
        }
}
```

（3）其余代码不变，同样可以实现 JdbcTemplate 相同的效果。

第 11 章 Spring 事务管理

11.1 事务的定义

事务就是指作为单个逻辑工作单元执行的一组数据操作序列，这些操作要么必须全部成功执行，要么必须全部执行失败，以保证数据的一致性和完整性。

事务具有 ACID 属性，具体解释起来是指事务具有原子性、一致性、隔离性和持久性四个主要特性。

原子性（Atomic）：事务由一个或多个行为绑在一起组成，好像是一个单独的工作单元。原子性确保在事务中的所有操作要么都发生，要么都不发生。

一致性（Consistent）：一旦一个事务结束了（不管成功与否），系统所处的状态和它的业务规则是一致的。即数据应当不会被破坏。

隔离性（Isolated）：事务应该允许多个用户操作同一个数据，一个用户的操作不会和其他用户的操作互相干扰。

持久性（Durable）：一旦事务完成，事务的结果应该持久化。

事务的 ACID 特性是由关系数据库管理系统（RDBMS）来实现的。

- 数据库管理系统采用日志来保证事务的原子性、一致性和持久性。日志记录了事务对数据库所做的更新，如果某个事务在执行过程中发生错误，就可以根据日志，撤销事务对数据库已做的更新，使数据库退回到执行事务前的初始状态。
- 数据库管理系统采用锁机制来实现事务的隔离性。当多个事务同时更新数据库相同的数据时，只允许持有锁的事务能更新该数据，其他事务必须等待，直到前一个事务释放了锁，其他事务才有机会更新该数据。

11.2 JDBC 数据库事务声明

数据库系统的客户程序只要向数据库系统声明了一个事务，数据库系统就会自动保证事务的 ACID 特性。在 JDBC API 中，java.sql.Connection 类代表一个数据库连接，它提供了以下方法控制事务。

- setAutoCommit(Boolean autoCommit)：设置是否自动提交事务；
- commit()：提交事务；
- rollback()：事务回滚。

JDBC API 声明事务的示例代码如下。具体的使用例子可以参看《基于 BootStrap3 的 JSP 项目实例教程》一书。

```
Connection = null;
    PreparedStatement pstmt = null;
    try{
    con = DriverManager.getConnection(dbUrl, username, password);
    //设置手工提交事务模式
    con.setAutoCommit(false);
    pstmt =......;
    pstmt.executeUpdate();
    //提交事务
    con.commit();
    }catch(Exception e){
    //事务回滚
    con.rollback();
    } finally{
       .......
    }
```

11.3 Spring 对事务管理的支持

Spring 框架提供了一致的事务管理抽象，这带来了以下好处：

- 为复杂的事务 API 提供了一致的编程模型，如 JTA、JDBC、Hibernate、JPA 和 JDO。
- 支持声明式事务管理。
- 提供比大多数复杂的事务 API（诸如 JTA）更简单的，更易于使用的编程式事务管理 API。
- 容易整合 Spring 的各种数据访问抽象。

Java EE 开发者有两个事务管理的选择：全局或本地事务。

- 全局事务：由应用服务器管理，使用 JTA。缺点是代码需要使用 JTA，使用笨重的 API。
- 本地事务：是和资源相关的。缺点是不能用于多个事务性资源。

Spring 解决了这些问题，它使应用开发者能够在任何环境下使用一致的编程模型。开发者可以只写一次自己的代码，这在不同环境下的不同事务管理策略中很有益处。Spring 框架同时提供声明式和编程式事务管理。多数情况下使用声明事务管理是多数使用者的首选。下面将分别讲解编程式事务管理和声明式事务管理。

11.3.1 Spring 编程式事务管理

Spring对事务控制的API全部位于org.springframework.transaction包下，除去异常定义的类，仅有 PlatformTransactionManager、SavepointManager、TransactionDefinition 和 TransactionStatus 四个接口。

- PlatformTransactionManager: 是一个事务管理平台，该接口有许多具体的事务实现类，例如 DataSourceTransactionManager, HibernateTransactionManager，等等。通过实现此接口，Spring 可以管理任何实现了这些接口的事务。开发人员也可以使用统一的编程模型来控制管理事务。
- SavepointManager: 事务回滚点管理接口，提供创建、释放回滚点，或者回滚到指定的回滚点。
- TransactionDefinition: 定义事务的名称、隔离级别、传播行为、超时时间长短、只读属性等。
- TransactionStatus: 获取事务的状态（回滚点、是否完成、是否新事物、是否回滚）属性，还可以进行事务 rollback-only 的设置。

下面将通过一个具体的案例来讲解 Spring 编程式事务管理。

（1）打开 Eclipse 开发环境，选择 Package Explorer，选中工程 spring8x0 并右击，在弹出的快捷菜单中选择“copy”，然后在 Package Explorer 的空白处右击，在弹出的菜单中选择“paste”，然后在弹出的对话框中的 Project name 输入框处输入工程名：spring10x0。

（2）修改 src 中 cap.dao.impl 子包中的 AdminDAOImpl 类的 addAdmin 方法，修改后的方法如下。

```
@Override
    public int addAdmin(Admin admin) {
        String sql="insert into admin(username,password) values(?,?)";
        int res=getJdbcTemplate().update(sql, new Object[]{admin.getUsername(),admin.getPassword()});
        int i=10/0;
        return res;
    }
```

代码解释:int i=10/0 此行代码由于分母为 0，所以会抛出 divide by zero 的异常，导致事务回滚。

（3）在工程 src 的 cap.service 子包中添加 AdminService 接口，编辑后的代码如下。

```
package cap.service;
import cap.bean.Admin;
public interface AdminService {
    public int addAdmin(Admin admin);
}
```

（4）在工程 src 的 cap.service.impl 中新建 AdminService 接口的实现类 AdminServiceImpl，编辑后的代码如下。

```
package cap.service.impl;
import org.springframework.transaction.PlatformTransactionManager;
import org.springframework.transaction.TransactionDefinition;
import org.springframework.transaction.TransactionStatus;
import cap.bean.Admin;
import cap.dao.AdminDAO;
import cap.service.AdminService;
public class AdminServiceImpl implements AdminService {
        private AdminDAO adminDAO;
        private TransactionDefinition txDefinition;
private PlatformTransactionManager txManager;
        public AdminDAO getAdminDAO() {
                return adminDAO;
        }
        public void setAdminDAO(AdminDAO adminDAO) {
                this.adminDAO = adminDAO;
        }
        public void setTxDefinition(TransactionDefinition txDefinition) {
        this.txDefinition = txDefinition;
        }
        public void setTxManager(PlatformTransactionManager txManager) {
            this.txManager = txManager;
        }
        @Override
        public int addAdmin(Admin admin) {
                int res=0;
                TransactionStatus status = txManager.getTransaction(txDefinition);
                try {
                        res=adminDAO.addAdmin(admin);
                        txManager.commit(status);
                } catch (Exception e) {
                        txManager.rollback(status);
        e.printStackTrace();
                }
                return res;
        }
}
```

（5）修改工程 src 中的 Spring 配置文件 applicationContext.xml，修改后的配置文件如下。

```
<?xml version="1.0" encoding="UTF-8"?>
<beans xmlns="http://www.springframework.org/schema/beans"
        xmlns:xsi="http://www.w3.org/2001/XMLSchema-instance" xmlns:context="http://www.
springframework.org/schema/context"
        xmlns:aop="http://www.springframework.org/schema/aop" xmlns:tx="http://www.springframework.
org/schema/tx"
        xsi:schemaLocation="
        http://www.springframework.org/schema/beans
```

```
        http://www.springframework.org/schema/beans/spring-beans.xsd
        http://www.springframework.org/schema/tx
        http://www.springframework.org/schema/tx/spring-tx.xsd
       http://www.springframework.org/schema/context
        http://www.springframework.org/schema/context/spring-context.xsd
        http://www.springframework.org/schema/aop
        http://www.springframework.org/schema/aop/spring-aop.xsd">
        <!-- 配置数据源 dataSource -->
        <bean id="dataSource"
              class="org.springframework.jdbc.datasource.DriverManagerDataSource">
              <property name="driverClassName" value="com.mysql.jdbc.Driver" />
              <property name="url" value="jdbc:mysql://localhost:3306/cap" />
              <property name="username" value="root" />
              <property name="password" value="admin" />
        </bean>
<!—定义事务管理器 txManager -->
        <bean id="txManager"
              class="org.springframework.jdbc.datasource.DataSourceTransactionManager">
              <property name="dataSource" ref="dataSource" />
        </bean>
              <!-- 配置 TransactionDefinition -->
        <bean id="txDefinition" class="org.springframework.transaction.support.DefaultTransactionDefinition">
 <property name="propagationBehaviorName" value="PROPAGATION_REQUIRED">
</property>
</bean>
<bean id="adminDAO" class="cap.dao.impl.AdminDAOImpl">
              <property name="dataSource" ref="dataSource"></property>
        </bean>
        <bean id="adminService" class="cap.service.impl.AdminServiceImpl">
              <property name="adminDAO" ref="adminDAO"></property>
              <property name="txManager" ref="txManager"></property>
              <property name="txDefinition" ref="txDefinition"></property>
        </bean>
</beans>
```

（6）修改 src 中 cap.test 包中 AdminDAOTest 类的 testAddAdmin 方法，修改后的代码如下。

```
@Test
    public void testAddAdmin() {
        AdminService adminService=(AdminService) ctx.getBean("adminService");
        Admin admin=new Admin();
        admin.setUsername("starlee2000");
        admin.setPassword("starlee2000");
        adminService.addAdmin(admin);
    }
```

（7）运行 testAddAdmin 方法，会出现如图 11-1 所示的结果。由于除数为零会导致事务回滚。

```
<terminated> AdminDAOTest.testAddAdmin (3) [JUnit] C:\Program Files\Java\jre7\bin\javaw.exe (2015年6月14日 下午9:26:22)
信息: Pre-instantiating singletons in org.springframework.beans.factory.support.Def
六月 14, 2015 9:26:23 下午 org.springframework.jdbc.datasource.DriverManagerDataSourc
信息: Loaded JDBC driver: com.mysql.jdbc.Driver
java.lang.ArithmeticException: / by zero
	at cap.dao.impl.AdminDAOImpl.addAdmin(AdminDAOImpl.java:54)
	at cap.service.impl.AdminServiceImpl.addAdmin(AdminServiceImpl.java:37)
	at cap.test.AdminDAOTest.testAddAdmin(AdminDAOTest.java:17)
```

图 11-1　工程 spring10x0 运行结果

11.3.2　Spring 事务管理

1. 使用 XML 声明事务管理

大多数 Spring 用户选择声明式事务管理。这是对应用代码影响最小的选择，因此也最符合非侵入式轻量级容器的理念。

Spring 在基于 XML（Schema）的配置中，添加了一个 tx 命名空间，在配置文件中以结构化的方式定义事务属性，大大提高了配置事务属性的便利性，配置 aop 的命名空间提供的切面定义，业务类方法事务的配置得到大大的简化。一个典型的基于 XML 声明事务配置代码如下。

```
<?xml version="1.0" encoding="UTF-8"?>
<beans xmlns="http://www.springframework.org/schema/beans"
      xmlns:xsi="http://www.w3.org/2001/XMLSchema-instance" xmlns:context="http://www.springframework.org/schema/context"
      xmlns:aop="http://www.springframework.org/schema/aop" xmlns:tx="http://www.springframework.org/schema/tx"
      xsi:schemaLocation="
      http://www.springframework.org/schema/beans
      http://www.springframework.org/schema/beans/spring-beans.xsd
      http://www.springframework.org/schema/tx
      http://www.springframework.org/schema/tx/spring-tx.xsd
     http://www.springframework.org/schema/context
         http://www.springframework.org/schema/context/spring-context.xsd
      http://www.springframework.org/schema/aop
      http://www.springframework.org/schema/aop/spring-aop.xsd">
  <!—代码①  -->
      <bean id="txManager"
          class="org.springframework.jdbc.datasource.DataSourceTransactionManager">
          <property name="dataSource" ref="dataSource" />
      </bean>
  <!—代码②  -->
      <tx:advice id="txAdvice" transaction-manager="txManager">
          <tx:attributes>
              <tx:method name="find*" read-only="true" />
              <tx:method name="*" />
          </tx:attributes>
      </tx:advice>
  <!—代码③  -->
      <aop:config>
```

```
        <aop:pointcut id="adminDAOPointCut" expression="execution(* cap.dao.impl.*.*(..))" />
        <aop:advisor advice-ref="txAdvice" pointcut-ref="adminDAOPointCut" />
    </aop:config>
</beans>
```

从上面的配置文件中，可以看见 Spring 配置文件中需要引入 tx 和 aop 的命名空间（加粗字体），代码①创建事务管理器 txManager，代码②使用<tx:advice>定义需要增强的方法，需要引用代码①创建的事务管理器 txManager。在<tx:advice>标签里面定义以 find 开头的方法不需要事务管理，余下的方法都需要事务管理。<tx: method>元素的属性如表 11-1 所示。代码③使用<aop:config>首先定义 adminDAOPointCut 切点，接着使用<aop:advisor>指定引用代码②创建的 txAdvice 和 adminDAOPointCut 切点。

表 11-1 <tx:method/>元素的属性

属性	必须否	默认值	描述
name	是		与事务属性关联的方法名。通配符（*）可以用来指定一批关联到相同的事务属性的方法。 如：'get*'、'handle*'、'on*Event'等等
propagation	否	REQUIRED	事务传播行为，可选值有 REQUIRED、SUPPOTS、NOT_SUPPORTED、NEVER 等
isolation	否	DEFAULT	事务隔离级别，可选的值有：DEFAULT、READ_UNCOMMITTED、READ_COMMITTED 等
timeout	否	-1	事务超时的时间（以秒为单位），如果设置为-1，事务超时的时间由底层的事务系统决定
read-only	否	false	事务是否只读
rollback-for	否		被触发进行回滚的 Exception(s)
no-rollback-for	否		不触发进行回滚的 Exception(s)

下面将通过一个实例讲解基于 XML 配置事务的使用。

（1）打开 Eclipse 开发环境，选择 Package Explorer，选中工程 spring10x0 并右击，在弹出的快捷菜单中选择“copy”，然后在 Package Explorer 的空白处右击，在弹出的菜单中选择“paste”，然后在弹出的对话框中的 Project name 输入框处输入工程名：spring10x1。

（2）修改 src 中 cap.service.impl 子包中的 AdminServiceImpl 类的 addAdmin 方法，修改后的方法如下。

```
@Override
    public int addAdmin(Admin admin) {
        return adminDAO.addAdmin(admin);
    }
```

（3）将上述的代码①～③添加到 Spring 的配置文件 applicationContext.xml，然后再添加下面的 Bean 配置。

```
<!-- 配置数据源 -->
    <bean id="dataSource"
        class="org.springframework.jdbc.datasource.DriverManagerDataSource">
        <property name="driverClassName" value="com.mysql.jdbc.Driver" />
```

```
            <property name="url" value="jdbc:mysql://localhost:3306/cap" />
            <property name="username" value="root" />
            <property name="password" value="admin" />
        </bean>
    <bean id="adminDAO" class="cap.dao.impl.AdminDAOImpl">
            <property name="dataSource" ref="dataSource"></property>
        </bean>
        <bean id="adminService" class="cap.service.impl.AdminServiceImpl">
            <property name="adminDAO" ref="adminDAO"></property>
        </bean>
```

（4）运行 cap.test 下的 testAddAdmin 测试方法，会出现和上一节相似的情况。

2. 使用注解实现事务管理

除了基于 XML 文件的声明式事务配置外，也可以采用基于注解式的事务配置方法。通过 @Transactional 注解对需要事务增强的 Bean 接口类或者方法进行标注，在容器中配置基于注解的事务增强驱动，就可以启用基于注解的声明式事务。

@Transactional 注解是用来指定接口、类或方法必须拥有事务语义的元数据。默认的 @Transactional 属性设置如下：

- 事务传播行为：PROPAGATION_REQUIRED。
- 事务隔离级别：ISOLATION_DEFAULT。
- 读写事务属性：读/写。
- 事务超时：默认是依赖于底层事务系统。
- 事务回滚：任何 RuntimeException 将触发事务回滚，但是任何 checked Exception 将不触发事务回滚。

这些默认的设置当然也是可以被改变的。@Transactional 注解的各种属性设置描述如表 11-2 所示。

表 11-2 @Transactional 注解的属性

属　性	类　型	描　述
传播性	枚举型：Propagation	可选的传播性设置
隔离性	枚举型：Isolation	可选的隔离性级别（默认值：ISOLATION_DEFAULT）
只读性	布尔型	读写型事务 vs. 只读型事务
超时	int 型（以秒为单位）	事务超时
回滚异常类（rollbackFor）	一组 Class 类的实例，必须是 Throwable 的子类	一组异常类，遇到时必须进行回滚。默认情况下 checked exceptions 不进行回滚，仅 unchecked exceptions（即 RuntimeException 的子类）才进行事务回滚
回滚异常类名（rollbackForClassname）	一组 Class 类的名字，必须是 Throwable 的子类	一组异常类名，遇到时必须进行回滚
不回滚异常类（noRollbackFor）	一组 Class 类的实例，必须是 Throwable 的子类	一组异常类，遇到时不回滚
不回滚异常类名（noRollbackForClassname）	一组 Class 类的名字，必须是 Throwable 的子类	一组异常类，遇到时不回滚

下面通过一个具体的案例讲解@Transactional 注解的使用。

（1）打开 Eclipse 开发环境，选择 Package Explorer，选中工程 spring10 并右击，在弹出的快捷菜单中选择“copy”，然后在 Package Explorer 的空白处右击，在弹出的菜单中选择“paste”，然后在弹出的对话框中的 Project name 输入框处输入工程名：spring11。

（2）在 src 里 cap.dao.impl 子包中的 AdminDAOImpl 类添加@Transactional，如图 11-2 所示。

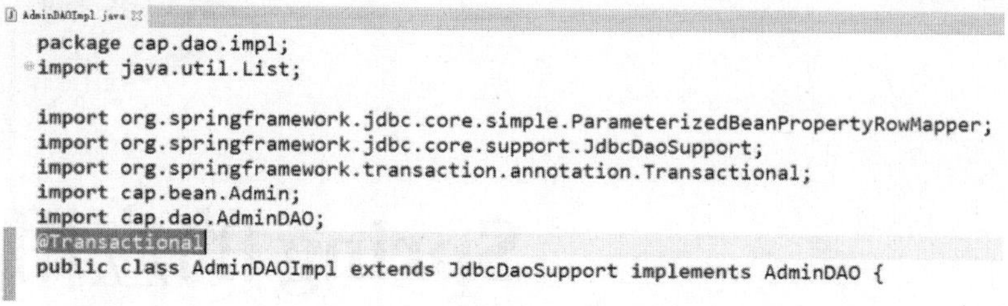

```
package cap.dao.impl;
import java.util.List;

import org.springframework.jdbc.core.simple.ParameterizedBeanPropertyRowMapper;
import org.springframework.jdbc.core.support.JdbcDaoSupport;
import org.springframework.transaction.annotation.Transactional;
import cap.bean.Admin;
import cap.dao.AdminDAO;
@Transactional
public class AdminDAOImpl extends JdbcDaoSupport implements AdminDAO {
```

图 11-2　@Transactional 的配置

（3）在 src 的 applicationContext.xml 中添加下面的配置，并且将<tx:advice>和<aop:config>标签注释。

```
<!--启动 spring 事务注解功能 -->
<tx:annotation-driven transaction-manager="transactionManager" />
```

（4）运行 cap.test 下的 testAddAdmin 测试方法，会出现和上一节相似的情况。

第12章 Spring的应用

本章主要讲解 Spring 和 Struts2 的整合步骤，接着讲解使用 Struts2+Spring+Spring JDBC 技术实现对一张表的“增、删、改、查”操作，然后在前一节的基础上添加分页显示的功能。

12.1 Struts2+Spring 实现增删改查

12.1.1 Struts2+Spring 整合

借助于 Spring 插件（Struts2-spring-plugin-XXX.jar），可以非常简单地完成 Spring 和 Struts2 的整合，这种整合包括让 Action 自动装配 Spring 容器中的 Bean，以及让 Spring 管理应用中的 Action。

下面讲述 Struts2+Spring 整合的四个关键点。

（1）在 web.xml 中添加 Struts2 的配置文件，添加的代码如下。

```
<filter>
        <filter-name>struts2</filter-name>
<filter-class>org.apache.struts2.dispatcher.ng.filter.StrutsPrepareAndExecuteFilter</filter-class>
    </filter>
    <filter-mapping>
        <filter-name>struts2</filter-name>
        <url-pattern>/*</url-pattern>
    </filter-mapping>
```

（2）Spring 提供一个 ContextLoaderListener 对象，该类可以作为 Web 应用的 Listener 使用，它会在 Web 应用启动时自动查找 WEB-INF/下的 Spring 的配置文件 applicationContext.xml，并且根据该文件来创建 Spring 容器。在 web.xml 文件中添加的配置代码如下。

```
<listener>
```

```
<listener-class>org.springframework.web.context.ContextLoaderListener</listener-class>
</listener>
<context-param>
    <param-name>contextConfigLocation</param-name>
    <param-value>classpath:/applicationContext.xml</param-value>
</context-param>
```

（3）使用 Spring 的容器管理 Action 类。

将 Struts2 的业务逻辑控制器类配置在 Spring 的配置文件中，业务逻辑控制器中引用的业务类一并注入。注意，必须将业务逻辑控制器 Bean 的作用域配置 scope="prototype"，示例如下。

```
<bean id="adminAction" class="cap.action.AdminAction" scope="prototype">
        <property name="adminService" ref="adminService"></property>
</bean>
```

在 struts.xml 或者 Struts2 配置文件中配置 Action 时，指定<action>的 class 属性为 Spring 配置文件中相应 Bean 的 id 或者 name 值。示例如下。

```
<action name="list" class="adminAction" method="list">
    <result name="success">/listAdmin.jsp</result>
</action>
```

（4）添加 Spring 插件（Struts2-spring-plugin-×××.jar）到工程的 WEB-INF 下的 lib 目录中，并配置 classpath。

12.1.2 Struts2+Spring 实现增删改查

下面通过具体的实例讲解 Struts2+Spring+SpringJDBC 实现“增、删、改、查”操作。

（1）在 Eclipse 中新建 Dynamic Web Project，工程名为 spring12，将 spring8 中 src 目录下的源码复制到 spring12，复制后的工程结构如图 12-1 所示。

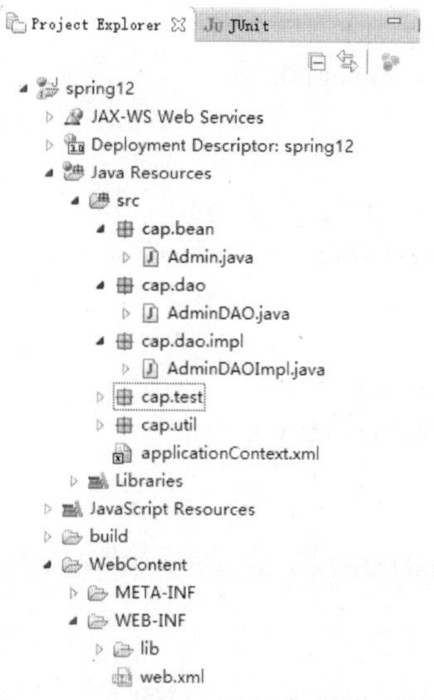

图 12-1 spring12 的工程结构图

（2）在 src 里的 cap.service 子包下新建 AdminService 接口，编辑后的代码如下。

```
package cap.service;
import java.util.List;
import cap.bean.Admin;
public interface AdminService {
        public List<Admin> findAdmins();
        public int updateAdmin(Admin admin);
        public int deleteAdmin(Integer id);
        public Admin findById(Integer id) ;
        public int addAdmin(Admin admin);
}
```

（3）在 src 中的 cap.service.impl 子包下新建 AdminService 接口的实现类 AdminServiceImpl，编辑后的代码如下。

```
package cap.service.impl;
import java.util.List;
import cap.bean.Admin;
import cap.dao.AdminDAO;
import cap.service.AdminService;
public class AdminServiceImpl implements AdminService {
        private AdminDAO adminDAO;
        @Override
        public List<Admin> findAdmins() {
                return adminDAO.findAdmins();
        }
        @Override
        public int updateAdmin(Admin admin) {
                return adminDAO.updateAdmin(admin);
        }
        @Override
        public int deleteAdmin(Integer id) {
                return adminDAO.deleteAdmin(id);
        }
        @Override
        public Admin findById(Integer id) {
                return adminDAO.findById(id);
        }
        @Override
        public int addAdmin(Admin admin) {
                return adminDAO.addAdmin(admin);
        }

        public AdminDAO getAdminDAO() {
                return adminDAO;
        }
        public void setAdminDAO(AdminDAO adminDAO) {
                this.adminDAO = adminDAO;
```

```
    }
}
```

（4）在 src 中的 cap.action 子包下新建 AdminAction 类，编辑后的代码如下。

```
package cap.action;
import java.util.List;
import cap.bean.Admin;
import cap.service.AdminService;
import com.opensymphony.xwork2.ActionSupport;
public class AdminAction extends ActionSupport{
    private List<Admin> adminList;
    private Integer id;
    private Admin admin;
    private AdminService adminService;
    public AdminService getAdminService() {
        return adminService;
    }
    public List<Admin> getAdminList() {
        return adminList;
    }
    public void setAdminList(List<Admin> adminList) {
        this.adminList = adminList;
    }
    public Integer getId() {
        return id;
    }
    public void setId(Integer id) {
        this.id = id;
    }
    public Admin getAdmin() {
        return admin;
    }
    public void setAdmin(Admin admin) {
        this.admin = admin;
    }
    public void setAdminService(AdminService adminService) {
        this.adminService = adminService;
    }
    public String list()
    {
        adminList=adminService.findAdmins();
        return SUCCESS;
    }
    public String delete()
    {
        int res=adminService.deleteAdmin(id);
        if(res>0)
            return SUCCESS;
```

```
            else
                    return ERROR;
        }
        public String add()
        {
            int res=adminService.addAdmin(admin);
            if(res>0)
                    return SUCCESS;
            else
                    return ERROR;
        }
        public String edit()
        {
            admin=adminService.findById(id);
            return SUCCESS;
        }
        public String update()
        {
            int res=adminService.updateAdmin(admin);
            if(res>0)
                    return SUCCESS;
            else
                    return ERROR;
        }
}
```

（5）在 src 的 spring 配置文件 applicationContext.xml 中添加下面代码。

```
<bean id="adminService" class="cdavtc.service.AdminServiceImpl">
<property name="adminDAO" ref="adminDAO"></property>
</bean>
<bean id="adminAction" class="cdavtc.action.AdminAction" scope="prototype">
<property name="adminService" ref="adminService"></property>
</bean>
```

（6）在 src 下新建 struts.xml，编辑后的代码如下。

```
<?xml version="1.0" encoding="UTF-8" ?>
<!DOCTYPE struts PUBLIC
        "-//Apache Software Foundation//DTD Struts Configuration 2.3//EN"
        "http://struts.apache.org/dtds/struts-2.3.dtd">
<struts>
        <package name="default" namespace="/" extends="struts-default">
                <action name="list" class="adminAction" method="list">
        <result name="success">/listAdmin.jsp</result>
        </action>
        <action name="del" class="adminAction" method="delete">
        <result name="success" type="redirect">list</result>
        <result name="error">/error.jsp</result>
```

```
        </action>
        <action name="edit" class="adminAction" method="edit">
        <result name="success" >/editAdmin.jsp</result>
        </action>
        <action name="update" class="adminAction" method="update">
        <result name="success" type="redirect">list</result>
        <result name="error">/error.jsp</result>
        </action>
        <action name="add" class="adminAction" method="add">
        <result name="success" type="redirect">list</result>
        <result name="error">/error.jsp</result>
        </action>
        </package>
    </struts>
```

（7）在 WebContent->WEB-INF 中的 web.xml 里添加下面的代码。

```
    <filter>
            <filter-name>struts2</filter-name>
        <filter-class>
    org.apache.struts2.dispatcher.ng.filter.StrutsPrepareAndExecuteFilter
    </filter-class>
        </filter>
        <filter-mapping>
            <filter-name>struts2</filter-name>
            <url-pattern>/*</url-pattern>
        </filter-mapping>
        <listener>
            <listener-class>
    org.springframework.web.context.ContextLoaderListener
    </listener-class>
        </listener>
        <context-param>
            <param-name>contextConfigLocation</param-name>
            <param-value>classpath:/applicationContext.xml</param-value>
        </context-param>
```

（8）在 WebContent 下新建 listAdmin.jsp 页面，代码如下。

```
    <%@ page language="java" contentType="text/html; charset=UTF-8"
        pageEncoding="UTF-8"%>
    <%@ taglib uri="/struts-tags" prefix="s"%>
    <!DOCTYPE html PUBLIC "-//W3C//DTD HTML 4.01 Transitional//EN" "http://www.w3.org/TR/html4/
loose.dtd">
    <html>
    <head>
    <meta http-equiv="Content-Type" content="text/html; charset=UTF-8">
    <title>显示所有用户</title>
    </head>
```

```
<body>
    <table>
        <s:iterator value="adminList" var="admin">
            <tr>
                <td><s:property value="#admin.id" /></td>
                <td><s:property value="#admin.username" /></td>
                <td><s:property value="#admin.password" /></td>
                <td><a href="del?id=<s:property value="#admin.id"/>">删除</a></td>
                <td><a href="edit?id=<s:property value="#admin.id"/>">编辑</a></td>
            </tr>
        </s:iterator>
    </table>
    <a href="addAdmin.jsp">添加用户</a>
</body>
</html>
```

（9）其余的页面参看第 5 章 struts2 应用中使用 Struts2+JDBC 实现“增、删、改、查”的资料。在地址栏上输入 http://localhost:8080/spring12/list，实现的结果如图 12-2 所示。

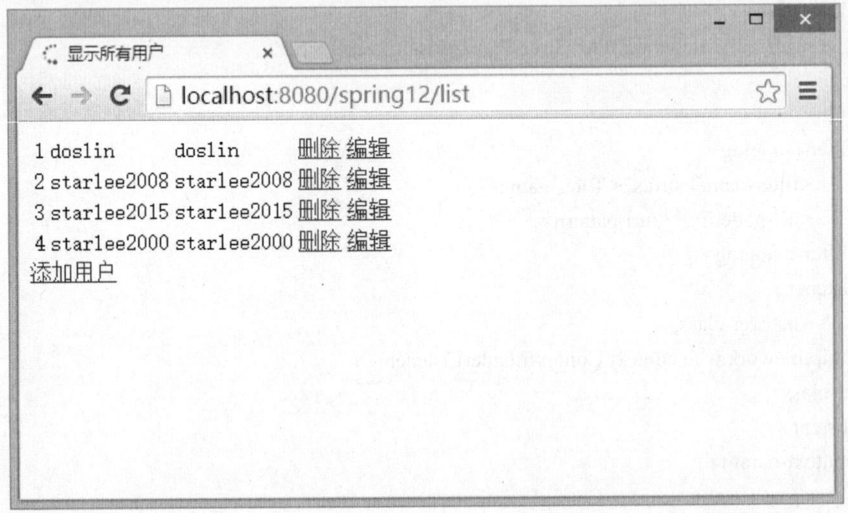

图 12-2　工程 spring12 的运行结果

12.2　Struts2+Spring 实现分页

Web 应用程序的一个重要实现功能是分页显示。数据库的数据量太大，不可能把几万行数据一次显示在浏览器上面，这样不现实（服务器和客服端都会出现负载过大）。实际情况是每页显示 10～20 行，是一个比较理想的显示状态。对于海量的数据查询，查询多少取多少，显然是最佳的解决办法。假如某个表中有 100 万条记录，第一页就取前 10 条，第二页取 11~20 条，在 MySQL 数据库中使用 limit 关键字实现，典型的分页 SQL 语句如下。

```
select * from table  limit  (currentPage - 1) * numPerPage, numPerPage;
```

下面将通过一个具体的案例讲解使用 Struts2+Spring+JDBC 技术实现分页查询。

（1）打开 Eclipse 开发环境，选择 Package Explorer，选中工程 spring12 并右击，在弹出的快捷菜单中选择"copy"，然后在 Package Explorer 的空白处右击，在弹出的菜单中选择"paste"，然后在弹出的对话框中的 Project name 输入框处输入工程名：spring13。

（2）在 src 中 cap.util 新建分页工具 PageBean 类，编辑后的代码如下。

```
package cap.util;
import java.util.List;
import org.springframework.jdbc.core.JdbcTemplate;
public class PageBean {
    private int totalPages;  // 总页数
    private int page;  // 当前页码
    @SuppressWarnings("rawtypes")
    private List resultList;  // 结果集存放 List
    @SuppressWarnings("rawtypes")
    public static   PageBean findByPage(String sql, int currentPage, int numPerPage,JdbcTemplate jTemplate){
        PageBean pageBean=new PageBean();
        //封装查询记录数的 sql 语句，本质是变成 select count(*) from table
        String countSQL = getSQLCount(sql);
        pageBean.setPage(currentPage);
        //计算总的记录数
        //Integer totalCount =jTemplate.queryForInt(countSQL);
        Integer totalCount=jTemplate.queryForObject(countSQL, Integer.class);
        pageBean.setTotalPages(numPerPage,totalCount);
        int startIndex = (currentPage - 1) * numPerPage;     //数据读取起始 index
        StringBuffer paginationSQL = new StringBuffer(" ");
        paginationSQL.append(sql);
        paginationSQL.append(" limit "+ startIndex+","+numPerPage);
        //根据分页 sql 语句查询
        List resultLists=jTemplate.queryForList(paginationSQL.toString());
        pageBean.setResultList(resultLists);
        return pageBean;
    }
    public static String getSQLCount(String sql){
        String sqlBak = sql.toLowerCase();
        String searchValue = " from ";
        String sqlCount = "select count(*) from "+ sql.substring(sqlBak.indexOf(searchValue)+searchValue.length(), sqlBak.length());
        return sqlCount;
    }
    public int getTotalPages() {
        return totalPages;
    }
    public void setTotalPages(int totalPages) {
        this.totalPages = totalPages;
    }
```

```
        public int getPage() {
                return page;
        }
        public void setPage(int page) {
                this.page = page;
        }
        public List getResultList() {
                return resultList;
        }
        public void setResultList(List resultList) {
                this.resultList = resultList;
        }
        // 计算总页数
        public void setTotalPages(int numPerPage,int totalRows) {
                if (totalRows % numPerPage == 0) {
                        this.totalPages = totalRows / numPerPage;
                } else {
                        this.totalPages = (totalRows / numPerPage) + 1;
                }
        }
}
```

（3）在 src 中 cap.dao 下的 AdminDAO 接口添加下面的代码。

```
public   PageBean findByPage(int currentPage, int numPerPage);
```

（4）在 src 中 cap.dao.impl 下的 AdminDAO 接口实现类 AdminDAOImpl，具体的实现代码如下。

```
@Override
        public PageBean findByPage(int currentPage, int numPerPage) {
                String sql="select * from admin";
return PageBean.findByPage(sql, currentPage,numPerPage,this.getJdbcTemplate());
        }
```

（5）在 src 中 cap.service 下的 AdminService 接口添加下面的代码。

```
public   PageBean findByPage(int currentPage, int numPerPage);
```

（6）在 src 中 cap.service.impl 下的 AdminService 接口实现类 AdminServiceImpl，具体的实现代码如下。

```
@Override
        public PageBean findByPage(int currentPage, int numPerPage) {
return adminDAO.findByPage(currentPage, numPerPage);
        }
```

（7）在 cap.action 子包中的 AdminAction 类，添加下面的代码。

```
private int page=1;
private PageBean pageBean;
//省略 getters 和 setter
```

```
public String listByPage()
    {
        pageBean=adminService.findByPage(page, 5);
        return SUCCESS;
    }
```

（8）在 src 的 struts2 配置文件 struts.xml 中添加下面的代码。

```
<action name="listByPage" class="adminAction" method="listByPage">
    <result name="success">/listAdminByPage.jsp</result>
</action>
```

（9）在 WebContent 下新建 listByPage.jsp 页面，编辑后的代码如下。

```
<%@ page language="java" contentType="text/html; charset=UTF-8"
    pageEncoding="UTF-8"%>
<%@ taglib uri="/struts-tags" prefix="s"%>
<!DOCTYPE html PUBLIC "-//W3C//DTD HTML 4.01 Transitional//EN" "http://www.w3.org/TR/html4/loose.dtd">
<html>
<head>
<meta http-equiv="Content-Type" content="text/html; charset=UTF-8">
<title>显示所有用户</title>
</head>
<body>
    <table>
        <s:iterator value="pageBean.resultList" var="admin">
            <tr>
                <td><s:property value="#admin.id" /></td>
                <td><s:property value="#admin.username" /></td>
                <td><s:property value="#admin.password" /></td>
                <td><a href="del?id=<s:property value="#admin.id"/>">删除</a></td>
                <td><a href="edit?id=<s:property value="#admin.id"/>">编辑</a></td>
            </tr>
        </s:iterator>
    </table>
    <!-- 分页条 -->
    <s:if test="%{pageBean.page <= 1}">
   [<font style="text-decoration: line-through">First</font>][<font
            style="text-decoration: line-through">Previous</font>]
   </s:if>
    <s:else>
   [<a href="listByPage.action?page=1">First</a>][<a
            href="listByPage.action?page=<s:property value="page-1"/>">Previous</a>]
    </s:else>
    <s:if test="%{pageBean.page >= pageBean.totalPages}">
    [<font style="text-decoration: line-through">Next</font>][<font
            style="text-decoration: line-through">Last</font>]
    </s:if>
```

```
        <s:else>
        [<a href="listByPage.action?page=<s:property value="page+1"/>">Next</a>][<a
                href="listByPage.action?page=<s:property value="pageBean.totalPages"/>">Last</a>]
        </s:else>
         page:
        <s:property value="pageBean.page" />
        /
        <s:property value="pageBean.totalPages" />
         
        <br />
        <a href="addAdmin.jsp">添加用户</a>
    </body>
    </html>
```

（10）运行工程，在地址栏中输入 http://localhost:8080/spring13/listByPage，实现的结果如图 12-3 所示。

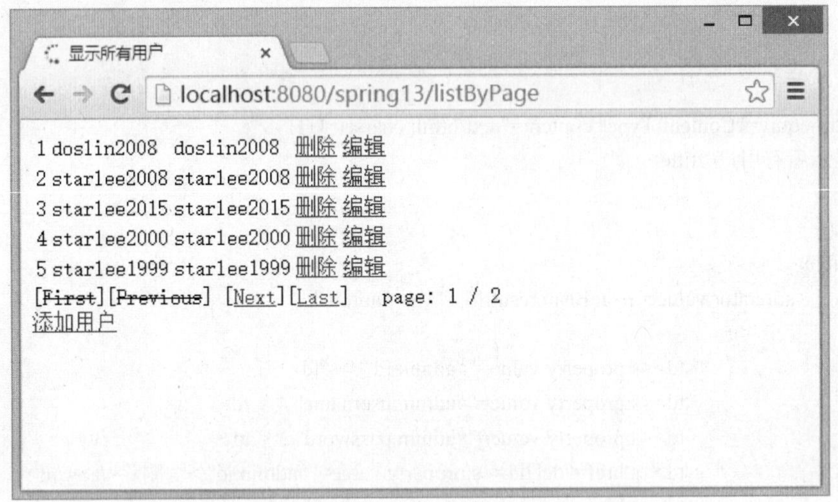

图 12-3 工程 spring13 的分页显示

第13章 Hibernate框架与入门

13.1 Hibernate 框架

13.1.1 ORM 概述

对象-关系映射英语简称为 ORM（Object-Relation Mapping），是用于将对象与对象之间的关系对应到数据库表与表之间的关系的一种模式。简单地说，ORM 是通过使用描述对象和数据库之间映射的元数据，将 Java 程序中的对象自动持久化到关系数据库中。对象和关系数据是业务实现的两种表现形式，业务实体在内存中表现为对象，在数据库中表现为关系数据。内存中的对象之间存在着关联和继承关系。而在数据库中，关系数据无法直接表达多对多关联和继承关系。因此，ORM 系统一般以中间件的形式存在，主要实现程序对象到关系数据库数据的映射。

一般的 ORM 包括四个部分：对持久类对象进行“增、删、改、查”操作的 API、用来规定类和类属性相关查询的语言或 API、规定 mapping metadata 的工具，以及可以让 ORM 实现同事务对象一起进行 dirty checking、lazy association fetching 和其他优化操作的技术。

具体说来，ORM 主要解决表 13-1 中的对应关系，即对象-关系的映射。

表 13-1 对象-关系对应关系

面向对象概念	面向关系概念
类	表
对象	表的行（记录）
属性	表的列（字段）
对象之间的关系	表与表之间的关系

ORM 的实现思想：将关系数据库表中的记录映射成为对象，以对象的形式呈现。程序员可以把对数据库的操作转化为对对象的操作。因此 ORM 的目的是为了方便开发人员以面向对象的思想来实现对数据库的操作。

ORM 采用元数据来描述对象-关系映射细节：元数据通常采用 XML 格式，并且存放在专门的对象-关系映射文件中。只要配置了持久化类与表的映射关系，ORM 框架在运行时就能够参照映射文件的信息，把域对象持久化到数据库中。常见的 ORM 框架有 Hibernate、MyBatis、TopLink 等。

13.1.2 Hibernate 简介

Hibernate 是一个开放源代码的对象关系映射框架，它对 JDBC 进行了非常轻量级的对象封装，使得 Java 程序员可以方便地使用对象编程思维来操纵数据库。 Hibernate 可以应用在任何使用 JDBC 的场合，既可以在 Java 的客户端程序使用，也可以在 Servlet/JSP 的 Web 应用中使用。

Hibernate 通过配置文件（hibernate.cfg.xml 或 hibernate.properties）和映射文件（*.hbm.xml）把 Java 对象或持久化对象（Persistent Object，PO）映射到数据库中的数据表，然后通过操作 PO，对数据库中的表进行各种操作，其中 PO 就是 POJO（Plain Old Java Objects，普通 Java 对象）加上映射文件（或映射注解）。Hibernate 的体系结构如图 13-1 所示。

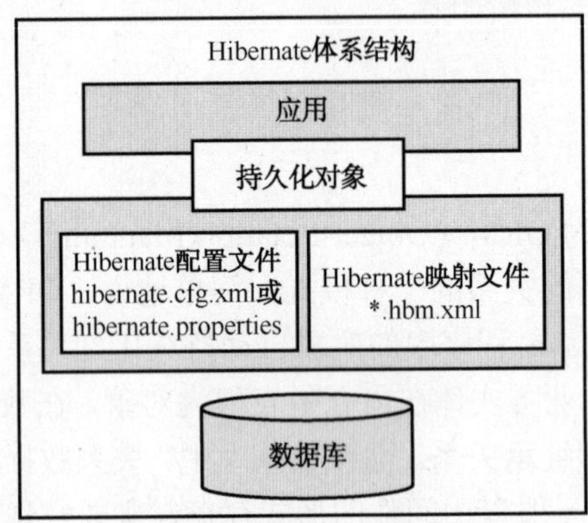

图 13-1　Hibernate 的体系结构

13.1.3 Hibernate 开发步骤

Hibernate 应用通常有两种开发方式：第一种是从持久化类到数据库表的开发方式，即先编写持久化类，然后手动编写或使用工具生成映射文件，进而生成数据库表的结构。第二种是从持久化类到数据库表的开发方式，即先创建数据库中表的结构，通过工具反向生成对应的映射文件和持久化类。

Hibernate 应用的开发大致可分为以下四个步骤：

（1）创建 Hibernate 的配置文件 hibernate.cfg.xml 或者 hibernate.properties。

（2）创建持久化类。

（3）创建对象-关系映射文件或者映射注解。

（4）通过 Hibernate API 编写访问数据库的代码。

本书主要采用从持久化类到数据库表的开发方式，这种方式符合面向对象的编程思路。实际开发中，可以根据情况选择对应的开发方式。

13.2　Hibernate 开发入门

13.2.1　搭建开发环境

本教材将使用 Eclipse 的 Hibernate tools 插件进行 Hibernate 应用开发。Eclipse 插件的安装方法详见附录 B。

本教材 Hibernate 应用采用的数据库为 product-order-cumstomer 数据库。要运行本案例，需要在 MySQL 数据库中创建 cdtu 数据库，并将数据的 Character set 设置为 UTF-8，避免中文出现乱码。

13.2.2　创建 Eclipse 工程

创建 Eclipse 工程的操作步骤如下。

（1）打开 Eclipse 开发环境，依次菜单选择“File→New”命令，新建 Java Project，工程名为 hibernate1。

（2）单击菜单“Window→Preference”，打开“Preferences”对话框，在左侧依次选择“Java→Build Path→User Libraries”，然后单击右侧的“New”按钮，在弹出对话框中输入 Hibernate4，单击“OK”按钮。如图 13-2 所示。

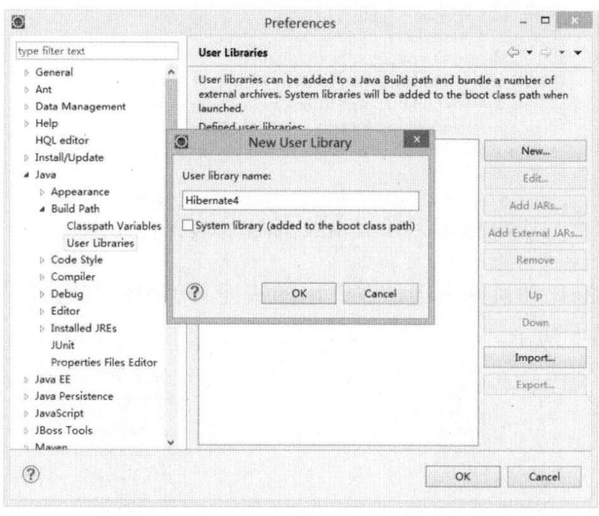

图 13-2　新建 User Library

（3）在图 13-2 中单击右侧的“Add JARs”按钮，将 jar 包添加到用户库 Hibernate4 中，单击“OK”按钮。继续在图 13-2 中单击“Add External JARs...”，在弹出的“JAR Selection”文件对话框选择添加的包，包括：Hibernate 必须的 jar 包，在 hibernate-release-4.3.8.Final\lib\required

目录下的所有 jar 包；C3P0 数据源的 jar 包，在 hibernate-release-4.3.8.Final\lib\optional\c3p0 目录下的所有 jar 包；MySQL 驱动 jar 包，mysql-connector-java-5.1.33-bin.jar。如图 13-3 所示。

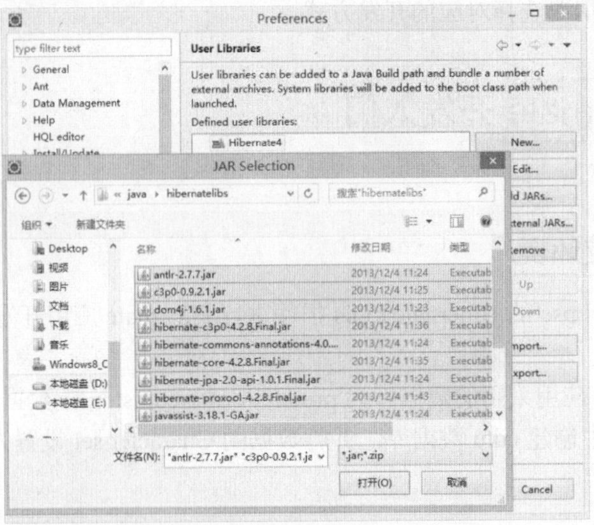

图 13-3　选中 User Library 所包含的 Jar 包

（4）添加完成后，用户库 Hibernate4 中包含如图 13-4 所示的 jar 包。

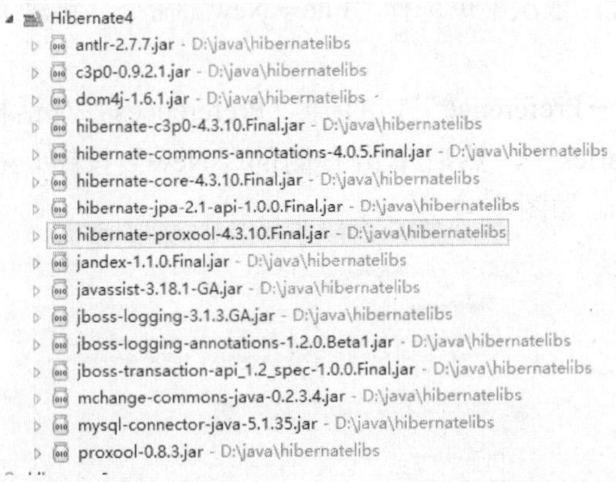

图 13-4　用户库 Hibernate4 中包含的 Jar 包

（5）选中 hibernate1 项目并右击，在弹出的快捷菜单中依次选择“Build Path→Configure Build Path”，如图 13-5 所示。

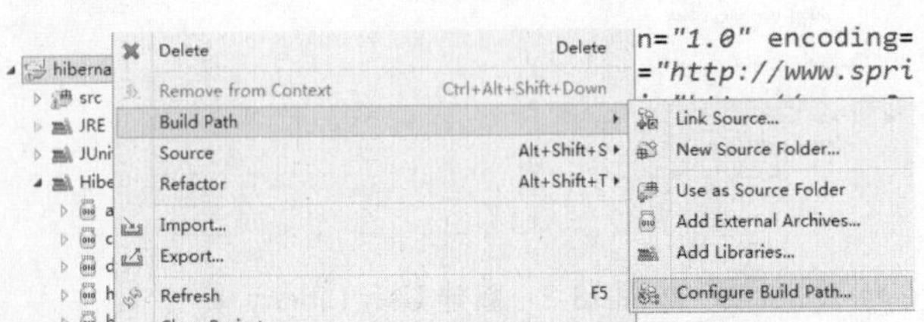

图 13-5　通过 Build Path 配置工程

（6）在弹出的对话框中选择 Libraries 页，单击“Add Library...”，在图 13-6 中选择 User

Library，单击“Next”按钮。然后勾选上一步创建的 Hibernate4，最后单击“Finish”按钮，如图 13-7 所示。

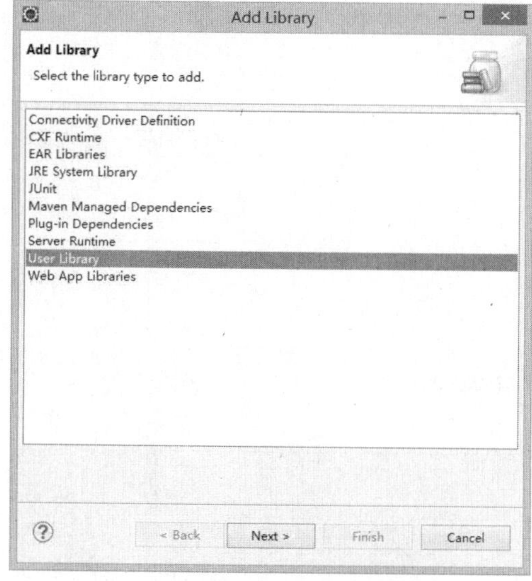

图 13-6 添加用户库

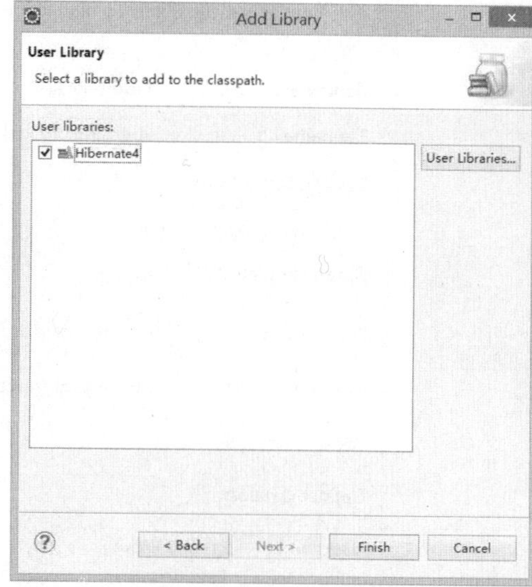

图 13-7 选中 Hibernate4 库

13.2.3 配置文件：hibernate.cfg.xml

（1）选中 src 并右击，在打开的快捷菜单中选择“New→Other”，出现“New”对话框，选择“Hibernate”选项，如图 13-8 所示。单击“Next”按钮，会出现“Create Hibernate Configuration File”对话框，如图 13-9 所示。生成的文件名和目录采用默认设置，单击“Next”按钮之后在弹出的对话框中输入如图 13-10 所示的内容，单击“Finish”按钮。

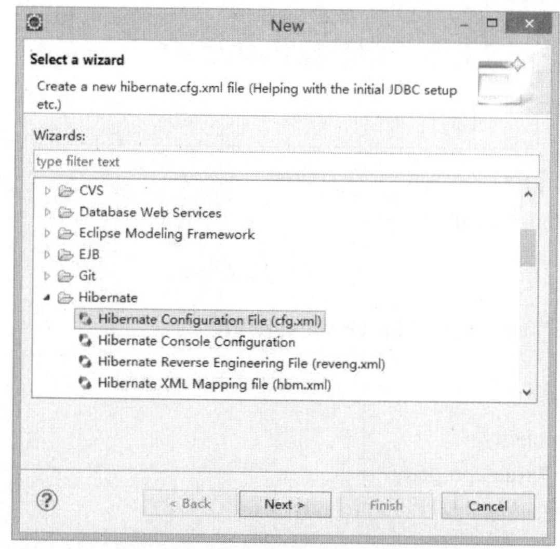

图 13-8 创建 Hibernate 配置文件

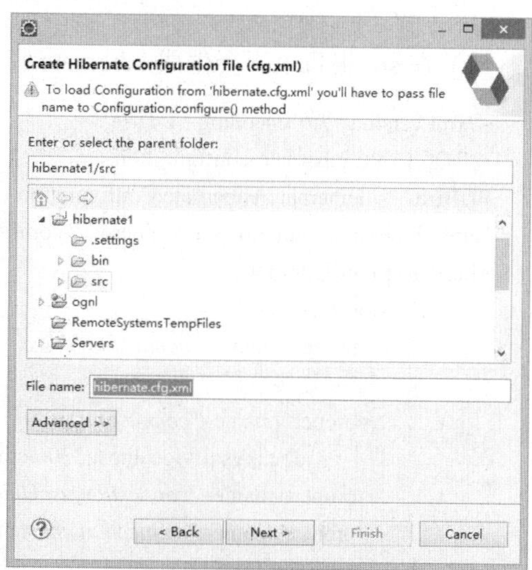

图 13-9 Hibernate 配置文件命名

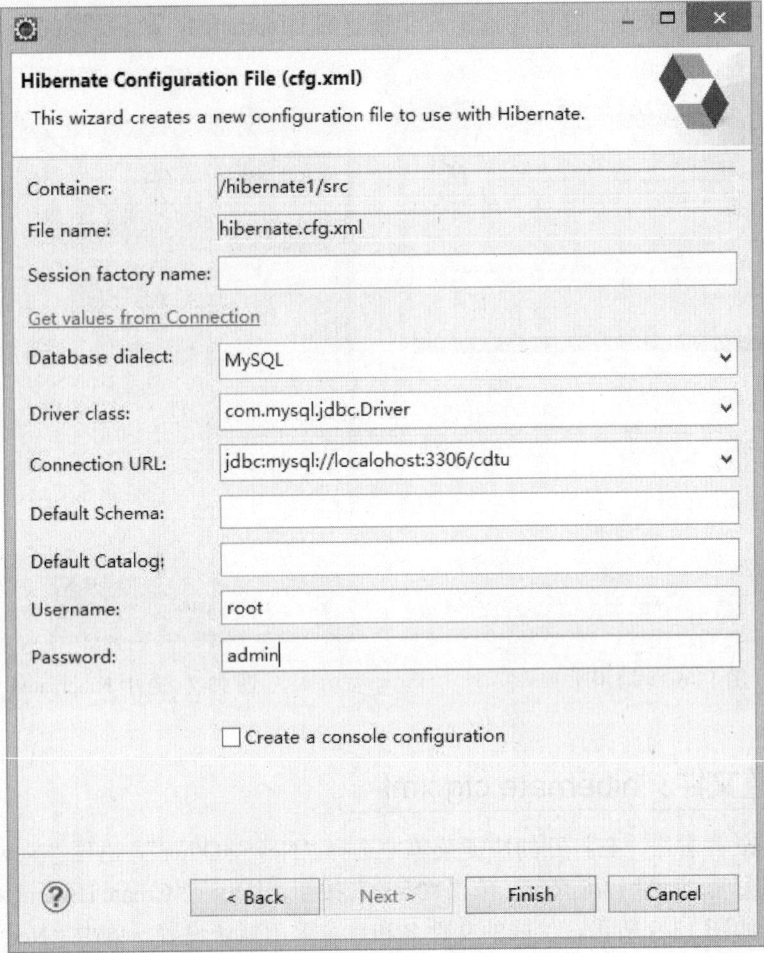

图 13-10　Hibernate 配置文件向导

（2）在 src 根目录下会生成 hibernate.cfg.xml 配置文件，编辑后的代码如下。

```
<?xml version='1.0' encoding='UTF-8'?>
<!DOCTYPE hibernate-configuration
PUBLIC "-//Hibernate/Hibernate Configuration DTD//EN"
"http://hibernate.sourceforge.net/hibernate-configuration-3.0.dtd">
<hibernate-configuration>
        <session-factory>
                <property name="connection.driver_class">com.mysql.jdbc.Driver</property>
                <!-- 连接数据库的 URL -->
                <property name="connection.url">
                        jdbc:mysql://localhost:3306/cdtu</property>
                <property name="connection.useUnicode">true</property>
                <property name="connection.characterEncoding">UTF-8</property>
                <!--连接的登录名 -->
                <property name="connection.username">root</property>
                <!--登录密码 -->
                <property name="connection.password">admin</property>
```

```
            <!-- C3P0 连接池设定 -->
            <property
name="hibernate.connection.provider_class">org.hibernate.connection.C3P0ConnectionProvider</property>
            <!--连接池的最小连接数 -->
            <property name="hibernate.c3p0.min_size">5</property>
            <!--最大连接数 -->
            <property name="hibernate.c3p0.max_size">30</property>
            <!--连接超时时间 -->
            <property name="hibernate.c3p0.timeout">1800</property>
            <!--statements 缓存大小 -->
            <property name="hibernate.c3p0.max_statements">100</property>
            <!--每隔多少秒检测连接是否可正常使用   -->
            <property name="hibernate.c3p0.idle_test_period">121</property>
            <!--当池中的连接耗尽的时候，一次性增加的连接数量,默认为 3 -->
            <property name="hibernate.c3p0.acquire_increment">1</property>
            <property name="hibernate.c3p0.validate">true</property>
            <!--是否将运行期生成的 SQL 输出到日志以供调试 -->
            <property name="show_sql">true</property>
            <!--指定连接的语言 -->
            <property name="dialect">org.hibernate.dialect.MySQLDialect</property>
            <!--映射 Product 这个资源 -->
            <mapping resource="cdtu/bean/Product.hbm.xml" />
        </session-factory>
    </hibernate-configuration>
```

代码解释：配置文件中的属性及其作用参看注释，在 13.3 节（第一个 Hibernate 应用详解）中会详细讲解配置文件中的属性及其作用。

13.2.4 创建持久化类 Product

在 src 的 cdtu.bean 中创建 Product 类，编辑后的代码如下。

```
package cdtu.bean;
public class Product {
    private Long id;// 产品编号
    private String name;// 产品名称
    private String serialNumber;// 产品序列号
    private double price;// 产品价格
    private int stock;// 产品库存量
    //省略 getters 和 setters 方法
}
```

13.2.5 创建对象-关系映射文件

接着为 Product 类创建对象-关系映射文件 Product.hbm.xml。选中 src 中的 cdtu.bean 并右击，依次选择“Other→Hibernate→Hibernate XML Mapping file”，如图 13-11 所示。

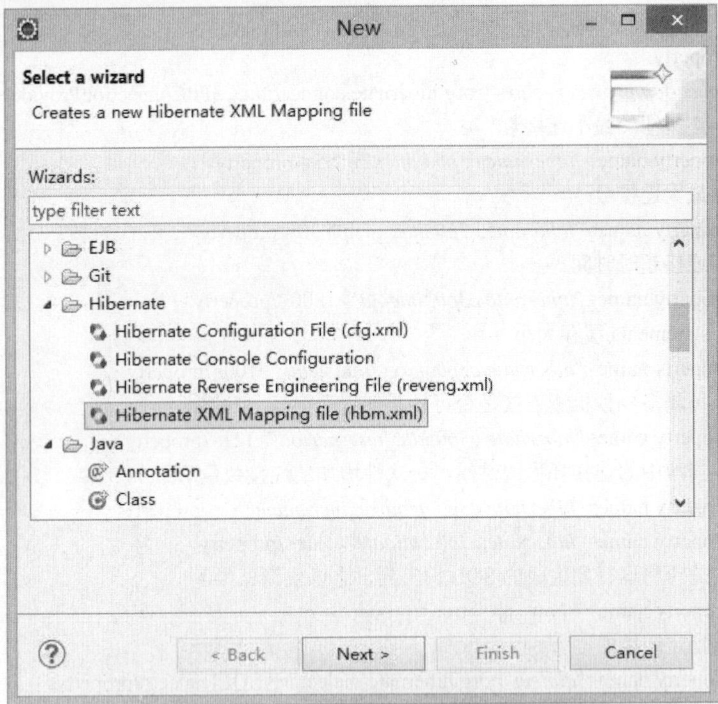

图 13-11　创建实体类的映射文件

Hibernate Tools 工具会自动生成配置文件 Product.hbm.xml，编辑后的代码如下。

```
<?xml version="1.0"?>
<!DOCTYPE hibernate-mapping PUBLIC "-//Hibernate/Hibernate Mapping DTD 3.0//EN"
"http://hibernate.sourceforge.net/hibernate-mapping-3.0.dtd">
<!-- Generated 2013-12-17 10:04:09 by Hibernate Tools 3.4.0.CR1 -->
<hibernate-mapping>
    <class name="cdtu.bean.Product" table="PRODUCT" >
        <id name="id" type="java.lang.Long">
            <column name="ID" />
            <generator class="increment" />
        </id>
        <property name="name" type="java.lang.String">
            <column name="NAME" />
        </property>
        <property name="serialnumber" type="java.lang.String">
            <column name="SERIALNUMBER" />
        </property>
        <property name="price" type="java.lang.Integer">
            <column name="PRICE" />
        </property>
        <property name="stock" type="java.lang.Integer">
            <column name="STOCK" />
        </property>
    </class>
```

```
</hibernate-mapping>
```

代码解释：<class>标签指定实体类 Product 映射的数据库表为 PRODUCT，<id>标签指定主键的生成方式，<property>标签指定了 Product 类的属性和数据库表 PRODUCT 的字段对应关系。

13.2.6 编写工具类

在 src 的 cap.util 子包中创建 HibernateSessionFactory.java 类，编辑后的代码如下。

```
public class HibernateUtils {
    private static String CONFIG_FILE_LOCATION = "/hibernate.cfg.xml";
    private static final ThreadLocal<Session> threadLocal = new ThreadLocal<Session>();
    private static Configuration configuration = new Configuration();
    private static SessionFactory sessionFactory;
    private static String configFile = CONFIG_FILE_LOCATION;
    /* 静态代码块创建 SessionFactory */
    static {
        try {
            configuration.configure(configFile);
            // 创建 SessionFactory
            StandardServiceRegistryBuilder srb = new StandardServiceRegistryBuilder().applySettings
(configuration.getProperties());
            StandardServiceRegistry sr = srb.build();
            sessionFactory = configuration.buildSessionFactory(sr);
        } catch (Exception e) {
            System.err.println("%%%% Error Creating SessionFactory %%%%");
            e.printStackTrace();
        }
    }
    private HibernateUtils() {}
    /**
     * 返回 ThreadLocal 中的 session 实例
     * @return Session
     * @throws HibernateException
     */
    public static Session getSession() throws HibernateException {
        Session session = (Session) threadLocal.get();
        if (session == null || !session.isOpen()) {
            if (sessionFactory == null) {
                rebuildSessionFactory();
            }
            session = (sessionFactory != null) ? sessionFactory.openSession() : null;
            threadLocal.set(session);
        }
        return session;
    }
    /**
```

```
         * 返回 Hibernate 的 SessionFactory
         */
        public static void rebuildSessionFactory() {
                try {
                        configuration.configure(configFile);
                        // 创建 SessionFactory
                        StandardServiceRegistryBuilder srb = new StandardServiceRegistryBuilder().applySettings
(configuration.getProperties());
                        StandardServiceRegistry sr = srb.build();
                        sessionFactory = configuration.buildSessionFactory(sr);
                } catch (Exception e) {
                        System.err.println("%%%% Error Creating SessionFactory %%%%");
                        e.printStackTrace();
                }
        }

        /**
         * 关闭 Session 实例并且把 ThreadLocal 中副本清除
         * @throws HibernateException
         */
        public static void closeSession() throws HibernateException {
                Session session = (Session) threadLocal.get();
                threadLocal.set(null);
                if (session != null) {
                        session.close();
                }
        }
        /**
         * 返回 SessionFactory
         */
        public static SessionFactory getSessionFactory() {
                return sessionFactory;
        }
        /**
         * return session factory
         * session factory will be rebuilded in the next call
         */
        public static void setConfigFile(String configFile) {
                HibernateUtils.configFile = configFile;
                sessionFactory = null;
        }
        /**
         * 返回 Configuration 对象
         */
        public static Configuration getConfiguration() {
                return configuration;
        }
}
```

本代码参考 MyEclipse 的 Hibernate 逆向工程自动生成的工具类。上述代码对 Hibernate 的常用操作进行了封装，提供了获取 SessionFactory、获取和关闭 Session 等操作的静态方法。

其中 java.lang.ThreadLocal 类可为每个线程保存一份独立的变量副本，所以每个线程都可以隔离访问自己的变量，而不会影响其他的线程。

13.2.7 编写测试用例

（1）选中工程 hibernate1 并右击 src，选择“New Junit Test Case”，在弹出的对话框中输入包名（Package）为 cdtu.test，类名（Name）为 HibernateTest，如图 13-12 所示。

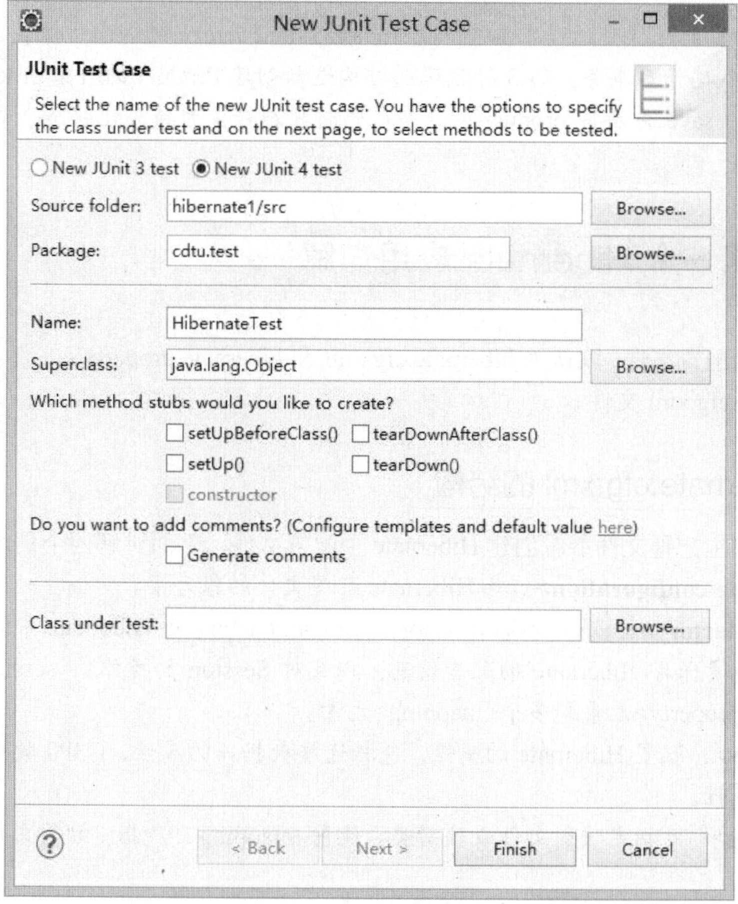

图 13-12 新建 Junit 测试类

（2）修改 cdtu.test 中的 HibernateTest.java 类，修改后的代码如下。

```
package cdtu.test;
import org.hibernate.Session;
import org.hibernate.Transaction;
import org.junit.Test;
import cdtu.bean.Product;
import cdtu.util.HibernateSessionFactory;
public class HibernateTest {
```

```
    private Session session;
    @Test
    public void testAdd() {
        session=HibernateSessionFactory.getSession();//第 1 行代码
        Transaction tx=session.beginTransaction();
        Product product=new Product("JSP 程序设计","123456789",19,20);
        session.save(product);
        tx.commit();
        session.close();
    }
}
```

代码解释：第 1 行代码通过工具类 HibernateSessionFactory 的静态方法获得 Session 对象 session，第 2 行代码开启事务，第 3 行代码通过构造器创建 Product 的对象 product，第 4 行代码通过 session 的 save 方法将 product 的状态从临时状态转变为持久化状态，最后一行代码关闭 session 对象。

13.3 第一个 Hibernate 应用详解

Hibernate 主配置文件：默认为 hibernate.cfg.xml 和 hibernate.properties 文件。在本教程中主要讲解 hibernate.cfg.xml 文件。

13.3.1 hibernate.cfg.xml 的结构

Hibernate 的主配置文件参看创建 Hibernate 主配置文件，下面讲解四个具体的标签。

- **<hibernate-configuration>**：是 Hibernate 配置文件的根元素。
- **<session-factory>**：配置 session-factory。SessionFactory 是 Hibernate 中的一个类，这个类主要负责保存 Hibernate 的配置信息，以及对 Session 的操作。<session-factory>可以有多个<property>元素和多个<mapping>元素。
- **<property>**：配置 Hibernate 的属性，包括连接数据库的属性、C3P0 数据库连接池属性和其他属性。
- **<mapping>**：可指定映射文件或映射类。使用<mapping .../>指定映射文件，示例代码如下：

```
<mapping resource="cdtu/bean/Productr.hbm.xml"/>
```

13.3.2 主要属性说明

使用<property>元素配置 Hibernate 的属性时，配置的 key 前面的 hibernate.前缀可以有，也可以没有。如 hibernate.dialect 或 dialect 都可以。但在与其他框架，如 Spring 一起使用时，最好加上 hibernate.前缀。Hibernate 的属性按作用分为以下几类：

（1）JDBC 连接属性，常用的属性如表 13-2 所示。

表 13-2 JDBC 常用的连接属性

属 性	含 义
connection.driver_class	指定数据库的 JDBC 驱动类。MySQL 的 JDBC 驱动类为： com.mysql.jdbc.Driver
connection.url	指定连接数据库的 URL。URL 的格式参考 JDBC，MySQL JDBC URL 格式如下：jdbc:mysql://[host:port]/[database][?参数名 1][=参数值 1][&参数名 2][=参数值 2]...
connection.username	指定连接数据库的用户名
connection.password	指定连接数据库的用户密码
connection.datasource	指定 JNDI 数据源的名字

（2）C3P0 数据库连接池属性，其常用的连接属性如表 13-3 所示。

表 13-3 C3P0 连接池常用属性

属 性	含 义
hibernate.connection.provider_class	该类用来向 Hibernate 提供 JDBC 连接
hibernate.c3p0.max_size	数据库连接池的最大连接数
hibernate.c3p0.min_size	数据库连接池的最小连接数
hibernate.c3p0.timeout	数据库连接池中连接对象超时未使用，就应该被销毁
hibernate.c3p0.max_statements	缓存 Statement 对象的数量
hibernate.c3p0.idle_test_period	表示连接池检测线程多长时间检测一次池内的所有链接对象是否超时
hibernate.c3p0.acquire_increment	当数据库连接池中的连接耗尽时，同一时刻获取多少个数据库连接
hibernate.c3p0.validate	是否每次连接都验证连接是否可用

（3）其他配置（<property ...>），其他配置的常用属性如表 13-4 所示。

表 13-4 其他配置的常用属性

属 性	含 义
dialect	配置数据库的方言，Hibernate 根据不同的底层数据库产生不同的 SQL 语句，Hibernate 会针对数据库的特性在访问时进行优化
show_sql	是否将运行期生成的 SQL 输出到日志以供调试。取值为 true 或 false
format_sql	是否将生成的 SQL 格式化，以便更好地调试程序。取值为 true 或 false
use_sql_comments	是否在 Hibernate 生成的 SQL 语句中添加有助于调试的注释
hbm2ddl.auto	在启动和停止时自动地创建，更新或删除数据库模式。取值为 create 、update、create-drop 和 validate。create：先删除，再创建表。update：如果表不存在就创建，不一样就更新。create-drop：初始化时创建，SessionFactory 执行 close()时删除。validate：验证表结构是否一致，如果不一致就抛出异常，在实际运行环境下使用
jdbc.fetch_size	实质是调用 Statement.setFetchSize()方法设定 JDBC 的 Statement，读取数据的时候每次从数据库中取出的记录条数
hibernate.jdbc.batch_size	设定对数据库进行批量删除，批量更新和批量插入时的批次大小，类似于设置缓冲区大小的意思。Batch Size 越大，批量操作向数据库发送 SQL 的次数越少，速度就越快

13.4 Hibernate 连接池

Hibernate 开发包提供了连接池的两种使用方式，上一节使用 C3P0 连接池技术，在本节中来讲解 proxool 连接池的使用，使用 proxool 连接池技术首先需要添加 proxool-0.x.x.jar 和 hibernate-proxool-4.x.x.Final.jar。下面接着讲解使用 proxool 配置 Hibernate 连接池的使用。

（1）打开 Eclipse 开发环境，选择 Package Explorer，选中工程 hibernate1 并右击，在弹出的快捷菜单中选择“copy”，然后在 Package Explorer 的空白处右击，在弹出菜单中选择“paste”，在 Project name 输入框中输入工程名: hibernate2。

（2）修改 Hibernate 的配置文件 hibernate.cfg.xml，代码如下。

```
<?xml version='1.0' encoding='UTF-8'?>
<!DOCTYPE hibernate-configuration
PUBLIC "-//Hibernate/Hibernate Configuration DTD//EN"
"http://hibernate.sourceforge.net/hibernate-configuration-3.0.dtd">
<hibernate-configuration>
    <session-factory>
        <property name="hibernate.connection.provider_class">
            org.hibernate.connection.ProxoolConnectionProvider
        </property>
        <property name="hibernate.proxool.pool_alias">DBPool</property>
        <property name="hibernate.proxool.xml">proxool.xml</property>
        <!--是否将运行期生成的 SQL 输出到日志以供调试 -->
        <property name="show_sql">true</property>
        <!--指定连接的语言 -->
        <property name="dialect">org.hibernate.dialect.MySQLDialect</property>
        <!--映射 Product 资源 -->
        <mapping resource="cdtu/bean/Product.hbm.xml" />
    </session-factory>
</hibernate-configuration>
```

属性解释：

- hibernate.connection.provider_class：定义 Hibernate 的连接加载类，Proxool 连接池是用 org.hibernate.connection.ProxoolConnectionProvider 类。
- hibernate.proxool.pool_alias ：连接池的别名。
- proxool.xml：声明连接池的配置文件位置，可以用相对或绝对路径。
- dialect：是声明使用数据的方言，在这里使用的是 MySQL 数据库。

（3）在 src 根目录中创建 proxool 的配置文件：proxool.xml，编辑后的代码如下。

```
<?xml version="1.0" encoding="UTF-8"?>
<!-- the proxool configuration can be embedded within your own application's.
    Anything outside the "proxool" tag is ignored. -->
<something-else-entirely>
    <proxool>
        <!--连接池的别名 -->
```

```
            <alias>DBPool</alias>
            <!--proxool 只能管理由自己产生的连接 -->
            <driver-url>
                  jdbc:mysql://localhost:3306/cdtu?useUnicode=true&amp;characterEncoding=UTF8
            </driver-url>
            <driver-class>com.mysql.jdbc.Driver</driver-class>
            <driver-properties>
                  <property name="user" value="root" />
                  <property name="password" value="admin" />
            </driver-properties>
            <!-- proxool 自动侦察各个连接状态的时间间隔(毫秒),侦察到空闲的连接就马上回收,超时的
            销毁 -->
            <house-keeping-sleep-time>90000</house-keeping-sleep-time>
            <!-- 指因未有空闲连接可以分配而在队列中等候的最大请求数,超过这个请求数的用户连接
            就不会被接受 -->
            <maximum-new-connections>20</maximum-new-connections>
            <!-- 最少保持的空闲连接数 -->
            <prototype-count>5</prototype-count>
            <!-- 允许最大连接数,超过了这个连接，再有请求时，就排在队列中等候，最大的等待请求数
由 maximum-new-connections 决定 -->
            <maximum-connection-count>100</maximum-connection-count>
            <!-- 最小连接数 -->
            <minimum-connection-count>10</minimum-connection-count>
      </proxool>
  </something-else-entirely>
```

代码解释：<driver-url>标签定义了连接到 MySQL 数据的 URL 地址，<driver-properties>标签定义了连接数据库需要的用户名和密码，其余标签的含义详见代码注释。

（4）运行工程中 cap.test 中测试类 HibernateTest 下面的 testAdd 方法，会实现和 13.2 节中相同的结果，向 MySQL 数据库中添加一条记录。

第14章

Hibernate 核心 API

Hibernate API 中提供了 Hibernate 的功能类和接口，应用程序通过这些类和接口可以直接以面向对象的方式访问数据库。Hibernate 的常用 API 如表 14-1 所示。

表 14-1　Hibernate 的常用 API

名　称	说　明
Configuration 类	用于配置、启动 Hibernate，创建 SessionFactory 实例对象
ServiceRegistry 接口	所有基于 Hibernate 的配置或者服务都必须统一向这个 ServiceRegistry 注册后才能生效
SessionFactory 接口	用于初始化 Hibernate，创建 Session 实例
Session 接口	用于保存、更新、删除、加载和查询持久化对象，充当持久化管理器
Transaction 接口	用于封装底层的事务，充当事务管理器

14.1　Hibernate 的运行过程

Hibernate 的运行过程大致可以分为以下的步骤：

（1）应用程序先调用 Configuration 类，该类读取 Hibernate 配置文件及映射文件中的信息。

（2）使用上述步骤这些信息生成一个 SessionFactory 对象。

（3）从 SessionFactory 对象生成一个 Session 对象。

使用 Session 对象生成 Transaction 对象，实现方式可分为以下两种：

① 可通过 Session 对象的 get()、load()、save()、update()、delete()和 saveOrUpdate()等方法对 PO 进行加载、保存、更新、删除等操作；

② 在查询的情况下，可通过 Session 对象生成一个 Query 对象，然后利用 Query 对象执行查询操作；如果没有异常，Transaction 对象将提交这些操作到数据库中。

14.2　Hibernate 核心 API

14.2.1　Configuration

Configuration 类负责管理 Hibernate 的配置信息。每个 Hibernate 配置文件对应一个 Configuration 对象。具体包括如下内容：

（1）Hibernate 运行的底层信息存储在 Hibernate 的主配置文件 hibernate.cfg.xml 中，包括连接数据库的 URL、用户名、密码、JDBC 驱动类，数据库方言（Dialect）、数据库连接池等信息。

（2）持久化类与数据表的映射关系（*.hbm.xml 文件）。

创建 Configuration 的两种方式

第一，属性文件（hibernate.properties）

```
Configuration cfg = new Configuration();
```

第二，xml 文件（hibernate.cfg.xml）

- 加载默认名称的配置文件（hibernate.cfg.xml）

```
Configuration cfg = new Configuration().configure();
```

- 加载指定名称的配置文件

```
Configuration cfg = new Configuration().configure("hibernate.cfg.xml");
```

14.2.2　ServiceRegistry

ServiceRegistry 接口，所有基于 Hibernate 的配置或者服务都必须统一向这个 ServiceRegistry 注册后才能生效。ServiceRegistry 对象是 Hibernate4 的配置入口，Configuration 对象将通过 ServiceRegistry 对象获取配置信息。使用了 Builder 模式创建一个 ServiceRegistry 对象。具体的使用代码如下。

```
StandardServiceRegistryBuilder srb = new StandardServiceRegistryBuilder().applySettings(configuration.
getProperties());
StandardServiceRegistry sr = srb.build();
```

14.2.3　SessionFactory

Configuration 对象根据当前的配置信息生成 SessionFactory 对象。SessionFactory 对象一旦构造完毕，即被赋予特定的配置信息（SessionFactory 对象中保存了当前的数据库配置信息和所有映射关系以及预定义的 SQL 语句。同时，SessionFactory 还负责维护 Hibernate 的二级缓存）。相关代码如下：

```
SessionFactory sessionFactory = configuration.buildSessionFactory(sr);
```

SessionFactory 常用的方法如表 14-2 所示。

表 14-2　SessionFactory 常用的方法

方　法	描　述
Session openSession()	打开 Session，需手动开发和关闭
Session getCurrentSession()	自动获取当前线程的 Session，若无则会新建。需在配置文件中配置 thread 属性，表明和当前线程绑定，无需手动关闭 session
void close()	关闭 Session

14.2.4 Session

Session 的常用 API

Session 是应用程序与数据库之间交互操作的一个单线程对象，是 Hibernate 运作的核心。所有持久化对象必须在 Session 的管理下才可以进行持久化操作。此对象的生命周期很短。Session 中有一个缓存，显式执行 flush()方法之前，所有的持久层操作的数据都缓存在 Session 对象处。一般一个业务操作使用一个 Session。持久化类与 Session 关联起来后就具有了持久化的能力。按照功能分，Session 接口方法分为获得持久化对象、操作对象、管理事务、管理 Session 四大类，具体的方法如表 14-3～表 14-6 所示。

表 14-3　获得持久化对象的方法

方　法	功　能
Object get(Class clazz, Serializable id)	根据给定标识（id）和实体类（clazz）返回持久化对象的实例，如果没有符合条件的持久化对象实例则返回 null
Object load（Class theClass, Serializable id)	在符合条件的实例存在的情况下，根据给定的实体类和标识返回持久化状态的实例。如果没有符合条件的持久化对象实例则会抛出异常

表 14-4　操作对象方法

方　法	功　能
Serializable save(Object object)	首先为给定的瞬时状态（Transient）的对象（根据配置）生成一个标识并赋值，然后将其持久化。返回生成的标识
void update(Object object)	根据给定的游离状态（detached）对象实例的标识更新对应的持久化实例
void delete(Object object)	从数据库中移除持久化（persistent）对象的实例
void saveOrUpdate(Object object)	根据给定的实例标识属性的值（注：可以指定 unsaved-value，一般默认为 null）来决定执行 save()或 update()操作
void persist(Object object)	将一个自由状态(transient)的实例持久化
Query createQuery(String queryString)	用给定的 HQL 查询字符串创建一个 Query 实例
Criteria createCriteria(Class persistentClass)	用给定的类（实体或子类/实现类）创建一个 Criteria 实例
SQLQuery createSQLQuery(String queryString)	用给定的 SQL 查询字符串创建一个 SQLQuery 实例

表 14-5　管理事务的方法

方　法	功　能
Transaction beginTransaction()	开始一个事务并且返回相关联的事务（Transaction）对象
Transaction getTransaction()	获取当前 Session 中关联的事务对象

表 14-6 管理 Session 的方法

方 法	功 能
boolean isOpen()	返回 session 是否打开
void flush()	强制提交清理（flush）Session。将 Session 缓存的数据写入数据库
void clear()	完全清除当前会话（Session）
void evict(Object object)	将当前对象实例从 session 缓存中清除
Connection close()	停止这个 Session，通过中断 JDBC 连接并且清空它

Session 中对象的状态

Hibernate 最核心的就是有关数据库的“增、删、改、查”操作，实现这些操作都需要依赖一个关键的对象 session。下面我们来看看 session 中对象的三个状态。

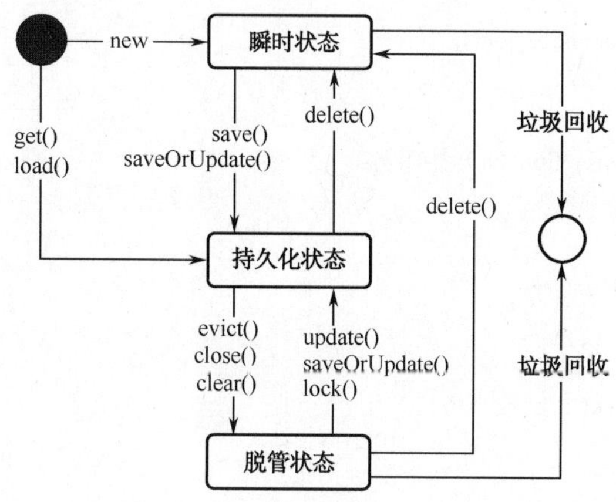

图 14-1 对象在 Session 中的状态

从图 14-1 可知，通常情况下，session 中的对象存在三种状态：瞬时（transient）、持久化（persistent）和脱管（detached）状态，还有一种移除（removed）状态未在图 14-1 中体现。

（1）瞬时状态（transient）：新创建的对象，没有和某个 session 进行关联，没有对象标识符(OID)。

（2）持久化状态（persistent）：与某个 session 进行关联，有对象标识符。数据库表中有对应的记录。session 在清理缓存时，会把此对象的数据与数据库表的数据进行同步。

（3）脱管状态（detached）：脱离了 session 的管理，有对象标识符。数据库表中有对应的记录，不保证此对象的数据与数据库表的数据是否同步。

（4）移除状态（removed）：与某个 session 进行关联。有对象标识符，数据库表中有对应的记录。session 在清理缓存时，会把数据库表对应的记录删除掉。这个对象不能再去使用它。

14.2.5 Transaction

Transaction 代表一次原子操作，它具有数据库事务的概念。所有持久层都应该在事务管理下进行，即使是只读操作。通过 session 的 beginTransaction()可获得 Transaction 的实例：

```
Transaction tx = session.beginTransaction();
```

Transaction 的常用方法如表 14-7 所示。

表 14-7　Transaction 的常用方法

方　法	功 能 说 明
void begin()	开始事务操作
void commit()	提交相关联的 session 实例
void rollback()	回滚（撤销）事务操作
boolean wasCommitted()	检查事务是否提交
boolean wasRolledBack()	检查事务是否回滚
void setTimeout(int seconds)	设置事务超时时间

一个事务可能包含多个持久化操作，应将它们放在开始事务和提交事务之间，从而形成一个完整的事务。在 Hibernate 中典型的事务处理代码如下。

```
Session session = factory.openSession();
Transaction tx;
try {
    tx = session.beginTransaction();//开始事务
    //多个持久化操作
    ...
    tx.commit();//提交事务
}
catch (Exception e) {
    if (tx!=null) tx.rollback();//回滚事务
    throw e;
}
finally {
    session.close();
}
```

14.3　Hibernate API 使用案例

本节利用上面提供的工具类讲解核心 API 的用法，并提供一个测试类，主要讲解使用 Hibernate API 实现对象之间的转化。

（1）在工程 hibernate1 的 cap.test 中新建 HibernateAPITest 测试类，并添加 testSessionCache 测试方法，编辑后的代码如下。

```
@Test
    public void testSessionCache() {
        Session session = HibernateSessionFactory.getSession();
        Transaction tx = session.beginTransaction();
        Product product1= (Product) session.get(Product.class, Long.valueOf(1));
        System.out.println(product1);
        Product product2= (Product) session.get(Product.class, Long.valueOf(1));
        System.out.println(product2);
        product2.setName("Lenovo");
```

```
        System.out.println(product2);
        /*flush()将数据库与缓存中的数据同步,不是必须调用的*/
        session.flush(); // 手动清理缓存
        /*
         * clear()写在 flush 后面，执行后才会引起缓存数据变化，
         */
        session.clear();
        product2.setPrice(4789);
        System.out.println(product2);
        tx.commit();
        session.close();
    }
```

代码解释：本测试方法主要讲解通过调用 session 的 get()方法将数据库对象转化到 session 中的持久化状态，在 session 中更改了对象的属性，然后通过 flush 方法将 session 中的对象和数据库对象同步到一致。

（2）在 HibernateAPITest 测试类中添加 testObjectStatus 测试方法，编辑后的代码如下。

```
@Test
    public void testObjectStatus() {
     Product product = new Product(); // transient 瞬时状态
     product.setName("Acer NoteBook");
        product.setSerialnumber("abcd879");
        product.setPrice(3498);
        Session session = HibernateSessionFactory.getSession();
        Transaction tx = session.beginTransaction();
        session.save(product); // persistent 持久化状态
        System.out.println(product);
        tx.commit();
        session.close();
        System.out.println(product); // detached 脱管
        Session session2 = HibernateSessionFactory.getSession();
        session2.beginTransaction();
        session2.save(product);
        System.out.println(product);// 脱管状态--> 持久化状态
        session2.getTransaction().commit();
        session2.close();
    }
```

代码解释：在本测试方法中，首先创建了一个 Product 对象，名为 product，由于没在 session 管理之中，所以这里对应的状态是瞬时状态；接着通过初始化 SessionFactory 并获得 session 对象，然后通过 session 的 save()方法将 product 对象保存到数据库，此时的 product 对应的状态是持久化状态；最后关闭 session，此时的 product 对象的状态转变为脱管状态。然后又重新开始打开 session，将 product 的状态转化为持久化状态。

（3）在 HibernateAPITest 测试类中添加 testGet 测试方法，编辑后的代码如下。

```
@Test
    public void testGet() {
```

```
    Session session = HibernateSessionFactory.getSession();
        session.beginTransaction();
        Product product = (Product) session.get(Product.class, Long.valueOf(1));
        System.out.println(product);
        session.getTransaction().commit();
        session.close();
    }
```

代码解释：在本测试方法中，首先打开 session，开启事务，然后通过 get()方法查找对象：首先查找 session 缓存中是否已经存在此标识符指定的对象，如果存在则直接使用，否则发出 SQL 语句从数据库中获取，如果数据库中也不存在，返回 null。

（4）在 HibernateAPITest 测试类中添加 testLoad 测试方法，编辑后的代码如下。

```
@Test
    public void testLoad() {
    Session session = HibernateSessionFactory.getSession();
        session.beginTransaction();
        Product product = (Product) session.load(Product.class, Long.valueOf(5));
        System.out.println(product.getId());
        System.out.println(product);
        session.getTransaction().commit();
        session.close();
    }
```

代码解释：load()方法先查找 session 缓存中是否已经存在此标识符指定的对象，如果存在，直接使用，否则 Hibernate 会为此标识符对象产生一个代理对象，实现延迟加载（懒加载）的功能。这个代理对象包含有 OID。当要使用到此对象的非 OID 属性值，才发出 SQL 语句去数据库中获取。如果数据库中也不存在，则返回 InvocationTargetException 异常。

（5）在 HibernateAPITest 测试类中添加 testDelete 测试方法，编辑后的代码如下。

```
@Test
    public void testDelete() {
    Session session = HibernateSessionFactory.getSession();
        session.beginTransaction();
        Product product = (Product) session.load(Product.class, Long.valueOf(12));
        session.delete(product);
        System.out.println(product);
        session.getTransaction().commit();
        session.close();
        System.out.println(product);
    }
```

代码解释：使用 delete()方法将对象从持久化状态转变到移除状态（处理移除状态的对象不要再去使用它，因为在 session 清理缓存时，数据库表中对应的数据会被删除掉）。

（6）在 HibernateAPITest 测试类中添加 testUpdate 测试方法，编辑后的代码如下。

```
@Test
public void testUpdate() {
    Session session = HibernateSessionFactory.getSession();
```

```
        session.beginTransaction();
        Product product = (Product) session.load(Product.class,Long.valueOf(8));
        product.setName("更新持久化状态的对象");
        session.getTransaction().commit();
        session.close();
        System.out.println(product);
        product.setName("修改脱管对象");
        Session session2 = HibernateSessionFactory.getSession();
        session2.beginTransaction();
        session2.update(product);
        session2.getTransaction().commit();
        session2.close();
    }
```

代码解释：在本例中，首先通过 load 获取一个持久化对象 product，然后修改其属性，通过提交事务更新到数据库中，session 关闭后 product 对象变为脱管对象，再次开启 session 通过 update 方法将脱管对象转化为持久化对象。

（7）在 HibernateAPITest 测试类中添加 testSaveOrUpdate 测试方法，编辑后的代码如下。

```
@Test
public void testSaveOrUpdate() {
        Session session = HibernateSessionFactory.getSession(); //第 1 行
        session.beginTransaction();
        Product product = (Product) session.get(Product.class, Long.valueOf(1));
        System.out.println(product);
        session.getTransaction().commit();
        session.close();  //第 4 行代码
        // 处理脱管状态
        product.setName("huawei");  //第 5 行代码
        Session session2 = HibernateSessionFactory.getSession();//第 6 行代码
        session2.beginTransaction();
        session2.saveOrUpdate(product);  //第 8 行代码
        System.out.println(product);
        session2.getTransaction().commit();
        session2.close();
        System.out.println(product);
    }
```

代码解释：第 1 行代码通过工具类 HibernateSessionFactory 的静态方法获得 session 对象，第 3 行代码调用 session 的 get 方法获得 OID 为 1 的对象 product，第 4 行代码关闭 session 之后，product 转变为脱管状态，第 5 行代码通过 setter 方法修改 product 对象的 name 属性，第 6 行代码重新获得 session2 对象，第 8 行代码调用 session2 的 saveOrUpdate 将 product 的脱管状态转变为持久化状态。

（8）在 HibernateAPITest 测试类中添加 testmerge 测试方法，编辑后的代码如下。

```
@Test
public void testmerge() {
        Product product = new Product();
```

```
// 对象没有与 session 关联，并且 OID 有值，就被认为是脱管对象
        product.setId(Long.valueOf(9));
        product.setName("Toshiba");
        Session session = HibernateSessionFactory.getSession();
        session.beginTransaction();
        session.get(Product.class, Long.valueOf(9));
        Product product1 = (Product) session.merge(product);
        session.getTransaction().commit();
        session.close();
    }
```

代码解释：merge()方法的使用情况：如果是瞬时对象，执行类似 save()的功能；如果是脱管对象，如果在当前 session 缓存中不存在同 OID 的对象，就执行类似 update()的功能；否则，把传入的对象数据合并到缓存中的对象，返回缓存中的对象。merge()方法也经常用来替代 update()和 saveOrUpdate()方法。

第15章 Hibernate映射

Hibernate 主要解决对象-关系的映射，具体包括类-表、对象-表的行（记录）、属性-表的列（字段）、对象之间的关系-表与表之间的关系的映射。

POJO 类和关系数据库之间的映射可以用一个 XML 文档来定义。通过 POJO 类的数据库映射文件，Hibernate 可以理解持久化类和数据表之间的对应关系，也可以理解持久化类属性与数据库表列之间的对应关系。在运行时，Hibernate 将根据这个映射文件来生成各种 SQL 语句。映射文件的扩展名为.hbm.xml。

15.1 Hibernate 映射文件结构

一个典型的 Hibernate 映射文件结构如下面的代码所示。

```
<hibernate-mapping>
    <class>
        <id>
            <column name=""></column>
            <type name="">
                <param name=""></param>
            </type>
            <generator class="">
                <param name=""></param>
            </generator>
        </id>
        <property >
            <column name=""></column>
            <type name=""></type>
        </property>
        <array name="">
```

```
            <key>
                <column name=""></column>
            </key>
            <index>
                <column name=""></column>
            </index>
            <element>
                <column name=""></column>
                <type name=""></type>
            </element>
        </array>
        <list name="">
            <key></key>
            <list-index></list-index>
            <one-to-many/>
        </list>
        <set name="">
            <key></key>
            <many-to-many>
                <column name=""></column>
            </many-to-many>
        </set>
        <map name="">
            <key></key>
            <map-key></map-key>
            <many-to-any id-type="">
                <meta-value class="" value=""/>
                <column name=""></column>
                <column name=""></column>
            </many-to-any>
        </map>
        <many-to-one name="">
            <column name=""></column>
        </many-to-one>
        <one-to-one name="">
            <formula></formula>
        </one-to-one>
        <component name="">
        </component>
    </class>
</hibernate-mapping>
```

映射文件的根元素是<hibernate-mapping>，该元素下可以拥有多个<class>子元素，每个class 元素对应一个持久化类的映射，即将类和表之间的关系进行映射。每个 Hibernate-mapping 中可以同时定义多个类，但建议为每个类都创建一个单独的映射文件。

根元素<hiberante-mapping>

<hiberante-mapping>元素包含很多属性，常用的属性如表 15-1 所示。

表 15-1 <hiberante-mapping>元素包的常用属性

属 性 名	功 能 说 明
package	指定一个包名，对于映射文件中非全限定的类名，默认在该包下
schema	指定映射数据库的 schema 名。若指定该属性，则会自动添加该 schema 前缀
catalog	指定映射数据库的 Catalog 名
default-cascade	设置 Hibernate 默认的级联风格，默认值是 none。若 Java 集合属性映射时没有指定 cascade 属性，则 Hibernate 将采用此处指定的级联风格
default-access	设置默认属性访问策略，默认值为 property，即使用 getter/setter 方法来访问属性。若指定 access，则 Hibernate 会忽略 getter/setter 方法，而通过反射访问成员变量（可以不定义 getter/setter 方法）
default-lazy	设置默认延迟加载策略，默认值为 true，即启用延迟加载策略。若 Java 属性映射时，集合映射时没有指定 lazy 属性，则 Hibernate 将采用此处指定的延迟加载策略
auto-import	是否允许在查询语言中使用非全限定的类名（仅限于本映射文件中的类），默认值为 true

15.2 类-表的映射

<class>元素用于指定类和表的映射。<class>元素的主要属性如表 15-2 所示。

表 15-2 <class>元素的主要属性

属 性 名	功 能 说 明
name	指定持久化类的类名
table	指定该持久化类映射的表名，默认以持久化类的类名作为表名
dynamic-insert	若设置为 true，表示当保存一个对象时，会动态生成 insert 语句，insert 语句中仅包含所有取值不为 null 的字段。默认值为 false
dynamic-update	若设置为 true，表示当更新一个对象时，会动态生成 update 语句，update 语句中仅包含所有取值需要更新的字段。默认值为 false
select-before-update	设置 Hibernate 在更新某个持久化对象之前是否需要先执行一次查询。默认值为 false。使用 select-before-update 通常会降低性能。如果重新连接一个脱管（detache）对象实例到一个 Session 中时，它可以防止数据库不必要触发 update
batch-size	指定根据 OID 来抓取实例时每批抓取的实例数。默认是 1
lazy	指定是否使用延迟加载
mutable	若设置为 true，等价于所有的<property>元素的 update 属性为 false，表示整个实例不能被更新。默认为 true
discriminator-value	指定区分不同子类的值。当使用<subclass/>元素来定义持久化类的继承关系时需要使用该属性

15.3 类的属性-数据表的字段的映射

属性的分类：标识属性、普通属性、集合属性、组件属性。

15.3.1 对象标识符映射

Hibernate 使用对象标识符（OID）来建立内存中的对象和数据库表中记录的对应关系，对象的 OID 和数据表的主键对应。Hibernate 通过标识符生成器来为主键赋值。

现代数据库都不推荐使用具有实际意义的物理主键，**推荐使用没有任何业务含义的逻辑主键**。代理主键通常为**整数类型**，因为整数类型比字符串类型要节省更多的数据库空间。

对象标识符的映射

在 Hibernate 映射文件中，<id>元素用来配置基本数据类型、基本类型的包装类、String、Date 等类型的对象标识符，定义持久化类的标识符属性（主键）。

<id>元素的主要属性如表 15-3 所示。

表 15-3 <id>元素的主要属性

属 性 名	功 能 说 明
name	标识持久化类 OID 的属性名
column	设置标识属性所映射的数据表的列名（主键字段的名字）
unsaved-value	若设定了该属性，Hibernate 会通过比较持久化类的 OID 值和该属性值来区分当前持久化类的对象是否为临时对象
type	指定 Hibernate 映射类型。Hibernate 映射类型是 Java 类型与 SQL 类型的桥梁。如果没有为某个属性显式设定映射类型，Hibernate 会运用反射机制先识别出持久化类的特定属性的 Java 类型，然后自动使用与之对应的默认的 Hibernate 映射类型

Java 的基本数据类型和包装器类型对应相同的 Hibernate 映射类型。基本数据类型无法表达 null，所以对于持久化类的 OID 推荐使用包装器类型。

<generator>元素用于设定持久化类标识符生成器。

<id>元素的<generator>子元素用来设置当前持久化类的标识符属性的生成策略。<generator>子元素主要包括一个 class 属性，用于指定使用的标识符生成器全限定类名或其缩写名。

主键生成策略 generator

Hibernate 提供了标识符生成器接口：IdentifierGenerator，并提供了各种内置实现。Hibernate 提供的内置标识符生成器如表 15-4 所示。

表 15-4 Hibernate 提供的内置标识符生成器

生 成 器	功 能 说 明
increment	由 Hibernate 获取数据库表中所有主键中的最大值，在最大值基础上加 1 的方式生成标识符。用于代理主键

续表

生 成 器	功 能 说 明
identity	由底层数据库以自动增长的方式生成标识符。MS SQL Server、MySQL 和 DB2 等数据库中可以设置表的某个字段（列）的数值自动增长。用于代理主键
sequence	由底层数据库的序列生成标识符。Oracle、DB2 等数据库可以创建一个序列，然后从序列中获取当前序号作为主键值。用于代理主键
hilo	由 Hibernate 以“high/low”高效算法生成 long、short 或 int 类型的标识符。用于代理主键
seqhilo	与 hilo 类似，但使用指定的 sequence 获取高位值。用于代理主键
uuid	由 Hibernate 用 128 位的 UUID 算法生成一个字符串类型的标识符。用于代理主键
guid	由 Hibernate 使用 GUID 字符串来生成标识符。用于代理主键
native	由 Hibernate 根据所使用的数据库支持能力自动选择 identity、sequence 或 hilo。用于代理主键
assigned	指派值，由 Java 应用程序负责生成标识符。用于自然主键
foreign	使用另外一个关联的持久化对象的标识符作为标识符。用于代理主键

15.3.2 普通属性映射

持久化类的普通属性是指基本数据类型、基本类型的包装器类、String、Date 等类型的属性。普通属性的值都可以直接保存到表的一个字段。

1. 普通属性的映射

在 Hibernate 中，<property>元素用于指定类的普通属性和表的字段的映射。

表 15-5 <property>元素的主要属性

属 性 名	功 能 说 明
name	指定该持久化类的属性的名字
type	指定 Hibernate 映射类型。Hibernate 映射类型是 Java 类型与 SQL 类型的桥梁。如果没有为某个属性显式设定映射类型，Hibernate 会运用反射机制先识别出持久化类的特定属性的 Java 类型，然后自动使用与之对应的默认的 Hibernate 映射类型
access	指定 Hibernate 的默认的属性访问策略。默认值为 property，即使用 getter、setter 方法来访问属性。若指定 field，则 Hibernate 会忽略 getter/setter 方法，而通过反射访问成员变量
update	数据库中该字段的值是否可以被修改
lazy	指定实例变量第一次被访问时，这个属性是否延迟抓取，默认值为 false
formula	设置一个 SQL 表达式，Hibernate 将根据它来计算出派生属性的值
column	指定与类的属性映射的表的字段名。如果没有设置该属性，将直接使用类的属性名作为字段名
unique	设置是否为该属性所映射的数据列添加唯一约束
not-null	若该属性值为 true，表明不允许为 null，默认为 false
index	指定一个字符串的索引名称。当系统需要 Hibernate 自动建表时，用于为该属性所映射的数据列创建索引，从而加快该数据列的查询
length	指定该属性所映射数据列的长度。默认值是 255

2. Java 类型、Hibernate 映射类型及 SQL 类型之间的对应关系

在 Hibernate 属性映射时，可以使用<property>元素的 type 属性指定 Hibernate 映射类型。Hibernate 映射类型是 Java 类型与 SQL 类型的桥梁。Hibernate 映射类型既可以是 Java 类型，也可以是 Hibernate 映射类型，具体类型参考表 15-6。如果没有为某个属性显式设定映射类型，Hibernate 会运用反射机制先识别出持久化类的特定属性的 Java 类型，然后自动使用与之对应的默认的 Hibernate 映射类型。

表 15- 6　Java 类型、Hibernate 映射类型及 SQL 类型之间的对应关系

Java 类型	Hibernate 映射类型	标准 SQL 类型	MySQL 类型
byte、java.lang.Byte	byte	TINYINT	TINYINT(4)
short、java.lang.Short	short	SMALLINT	SMALLINT(6)
int、java.lang.Integer	int、integer	INGEGER	INT(11)
long、java.lang.Long	long	BIGINT	BIGINT(20)
float、java.lang.Float	float	FLOAT	FLOAT
double、java.lang.Double	double	DOUBLE	DOUBLE
java.math.BigDecimal	big_decimal	NUMERIC	DECIMAL(19,2)
java.math.BigDecimal		NUMERIC	DECIMAL(19,2)
boolean、java.lang.Boolean	boolean	BIT	BIT(1)
boolean、java.lang.Boolean	yes_no	CHAR(1)('Y'或'N')	CHAR(1)
boolean、java.lang.Boolean	true_false	CHAR(1)('T'或'F')	CHAR(1)
char、java.lang.Character	character	CHAR(1)	CHAR(1)
java.lang.String	string	VARCHAR	VARCHAR(255)
java.util.Date、java.sql.Date	date	DATE	DATE
java.util.Date、java.sql.Time	time	TIME	TIME
java.util.Date、java.sql.Timestamp	timestamp	TIMESTAMP	TIMESTAMP
java.util.Calendar	calendar	TIMESTAMP	TIMESTAMP
java.util.Calendar	calendar_date	DATE	DATE
byte[]	binary	VARBINARY、BLOB	
java.lang.String	text	CLOB	TEXT
java.io.Serializable	serializable	VARBINARY、BLOB	
java.sql.Clob	clob	CLOB	
java.sql.Blob	blob	BLOB	
java.lang.Class	class	VARCHAR	VARCHAR(255)
java.util.Locale	locale	VARCHAR	VARCHAR(255)
java.util.TimeZone	timezone	VARCHAR	VARCHAR(255)
java.util.Currency	currency	VARCHAR	VARCHAR(255)

注：在 MySQL 类型中如果为空，说明 MySQL 没有对应的数据类型。

下面接着通过一个具体的实例讲解 Java 类型、Hibernate 映射类型与 SQL 类型之间的关系。

（1）在 Eclipse 开发环境中，创建 Java Project，工程名为 hibernate3。

（2）在 src 中的 cap.bean 中创建持久化类 CommonAttributeDemo.java，编辑后的代码如下。

```
package cdtu.bean;
public class CommonAttributeDemo {
	private Integer id;
	private Byte byteAttr;
	private short shortAttr;
	private int intAttr;
	private Long longAttr;
	private float floatAttr;
	private Double doubleAttr;
	private java.math.BigDecimal bigDecimalAttr;
	private java.math.BigDecimal bigDecimalAttr2;
	private boolean booleanAttr_boolean;
	private boolean booleanAttr_yes_no;
	private boolean booleanAttr_true_false;
	private char charAttr;
	private String stringAttr;
	private String stringAttr2;
	private Class classAttr_class;
	private java.util.Locale localeAttr;
	private java.util.TimeZone timeZoneAttr;
	private java.util.Currency currencyAttr;

	//省略 getters 和 setters
}
```

（3）使用 Hibernate Tools 在 cap.bean 下创建 CommonAttributeDemo 类的映射文件，编辑后的代码如下。

```
<?xml version="1.0"?>
<!DOCTYPE hibernate-mapping PUBLIC "-//Hibernate/Hibernate Mapping DTD 3.0//EN"
"http://hibernate.sourceforge.net/hibernate-mapping-3.0.dtd">
<hibernate-mapping>
	<class name="cdtu.bean.CommonAttributeDemo" table="COMMONATTRIBUTEDEMO">
		<id name="id" type="java.lang.Integer">
			<column name="ID" />
			<generator class="native" />
		</id>
		<property name="byteAttr" type="java.lang.Byte">
			<column name="BYTEATTR" />
		</property>
		<property name="shortAttr" type="short">
			<column name="SHORTATTR" />
		</property>
		<property name="intAttr" type="int">
			<column name="INTATTR" />
		</property>
```

```
        <property name="longAttr" type="java.lang.Long">
            <column name="LONGATTR" />
        </property>
        <property name="floatAttr" type="float">
            <column name="FLOATATTR" />
        </property>
        <property name="doubleAttr" type="java.lang.Double">
            <column name="DOUBLEATTR" />
        </property>
        <property name="bigDecimalAttr" type="java.math.BigDecimal">
            <column name="BIGDECIMALATTR" />
        </property>
        <property name="bigDecimalAttr2" type="java.math.BigDecimal">
            <column name="BIGDECIMALATTR2" precision="12" scale="4" />
        </property>
        <property name="booleanAttr_boolean" type="boolean">
            <column name="BOOLEANATTR_BOOLEAN" />
        </property>
        <property name="booleanAttr_yes_no" type="yes_no">
            <column name="BOOLEANATTR_YES_NO" />
        </property>
        <property name="booleanAttr_true_false" type="true_false">
            <column name="BOOLEANATTR_TRUE_FALSE" />
        </property>
        <property name="charAttr" type="char">
            <column name="CHARATTR" />
        </property>
        <property name="stringAttr" type="java.lang.String">
            <column name="STRINGATTR" />
        </property>
        <property name="stringAttr2" type="java.lang.String">
            <column name="STRINGATTR2" length="20" />
        </property>
        <property name="classAttr_class" type="java.lang.Class">
            <column name="CLASSATTR_CLASS" />
        </property>
        <property name="localeAttr" type="java.util.Locale">
            <column name="LOCALEATTR" />
        </property>
        <property name="timeZoneAttr" type="java.util.TimeZone">
            <column name="TIMEZONEATTR" />
        </property>
        <property name="currencyAttr" type="java.util.Currency">
            <column name="CURRENCYATTR" />
        </property>
    </class>
</hibernate-mapping>
```

代码解释：<class>标签中定义了 cdtu.bean 子包中 CommonAttributeDemo 持久化类对应的数据库表为 COMMONATTRIBUTEDEMO，<id>标签定义了主键自动生成方式为 native。其余的 <property> 标签定义了持久化类 CommonAttributeDemo 中的属性和数据库中 COMMONATTRIBUTEDEMO 表字段的对应关系。

（4）复制 hibernate1 工程中的配置文件 hibernate.cfg.xml 到 src 根目录下，在其中添加下面的属性并将资源的映射文件修改如下：

```
<property name="hbm2ddl.auto">update</property>
<!--映射资源 -->
<mapping resource="cdtu/bean/CommonAttributeDemo.hbm.xml" />
```

代码解释：添加 hbm2ddl.auto 属性为 update 是为了自动在数据库中生成需要的表，如果表已经存在，但与已存在的表结构不一致，将会更新表结构。

（5）在 cap.test 子包中创建 CommonAttribute 测试类，在测试类中添加如下的测试方法。

```
@Test
    public void testCommonAttribute() {
        Session sess = HibernateSessionFactory.getSession();
        Transaction tx = null;
        CommonAttributeDemo attrDemo = new CommonAttributeDemo();//第 3 行代码
        attrDemo.setByteAttr((byte)255);
        attrDemo.setShortAttr((short)256);
        attrDemo.setIntAttr(100);
        attrDemo.setLongAttr(1000L);
        attrDemo.setFloatAttr(1234.56789F);
        attrDemo.setDoubleAttr(1234.56789);
        attrDemo.setBigDecimalAttr(new java.math.BigDecimal(1234.56789));
        attrDemo.setBooleanAttr_boolean(true);
        attrDemo.setBooleanAttr_yes_no(true);
        attrDemo.setBooleanAttr_true_false(true);
        attrDemo.setCharAttr('a');
        attrDemo.setStringAttr("Hello World");
        attrDemo.setStringAttr2("Hello World 2");
        attrDemo.setClassAttr_class((Class)(cdtu.bean.CommonAttributeDemo.class));
        attrDemo.setLocaleAttr(java.util.Locale.CHINA);
        attrDemo.setTimeZoneAttr(java.util.TimeZone.getDefault());
        //下面一行为第 20 行代码
        attrDemo.setCurrencyAttr(java.util.Currency.getInstance(java.util.Locale.US));
        try {
            tx = sess.beginTransaction();
            sess.save(attrDemo);
            tx.commit();
        } catch (Exception e) {
            if (tx != null)
                tx.rollback();
            throw e;
        } finally {
            HibernateSessionFactory.closeSession();
```

```
            }
        }
    }
```

代码解释：第 3 行代码创建了 CommonAttributeDemo 的对象 attrDemo，第 4～20 行代码通过 setter 方法设置属性，余下的代码首先开启事务，通过 session 的 save 方法将对象 attrDemo 转变为持久化状态，并提交事务，如果发生异常，则事务回滚，撤销前面的数据库插入操作。最后关闭 session 对象。

（6）打开 Navicat，在数据库 cdtu 中查看生成的表结构，同时添加了测试用例中的数据，请读者自行打开数据库表进行查看。

映射属性生成的表结构如图 15-1 所示。

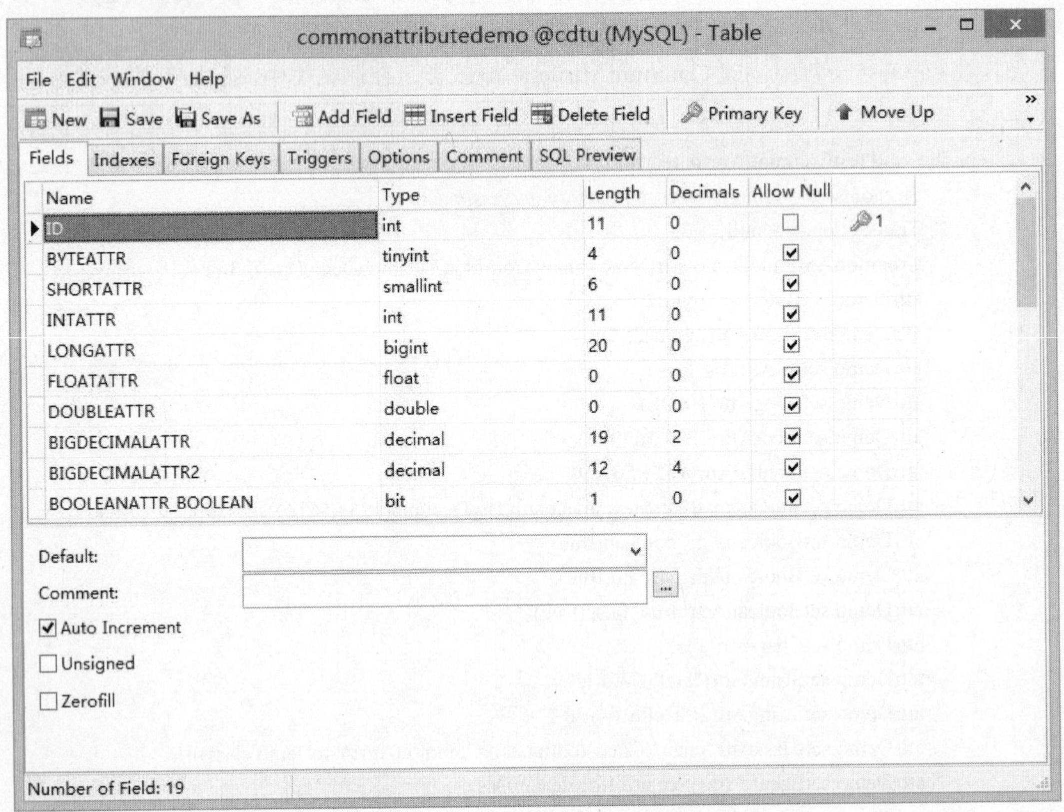

图 15-1　映射属性生成的表结构

3. 日期、时间类型属性的映射

在 Java 中，代表时间和日期的类型包括：java.util.Date 和 java.util.Calendar。此外，在 JDBC API 中还提供了 3 个扩展了 java.util.Date 类的子类：java.sql.Date、java.sql.Time 和 java.sql.Timestamp，这三个类分别和标准 SQL 类型中的 DATE、TIME 和 TIMESTAMP 类型对应。

在标准 SQL 中，DATE 类型表示日期，TIME 类型表示时间，TIMESTAMP 类型表示时间戳，同时包含日期和时间信息。

表 15-7 是 Hibernate 日期类型的内置映射方式。

表 15-7 Hibernate 日期类型的内置映射方式

Hibernate 映射类型	Java 类型	标准 SQL	取值
timestamp	java.util.Date java.sql.Timestamp	TIMESTAMP	表示时间和日期： YYYYMMDDhhmmss
calendar	java.util.Calendar	TIMESTAMP	表示时间和日期： YYYYMMDDHHMMSS
date	java.util.Date java.sql.Date	DATE	表示日期： YYYY-MM-DD
calendar_date	java.util.Calendar	DATE	表示日期： YYYY-MM-DD
time	java.util.Date java.sql.Time	TIME	表示时间： hh:mm:ss

下面接着通过具体的实例讲解日期、时间类型属性的映射方式。

（1）继续在工程“hibernate3→src→cap.bean”下创建 DateTimeDemo.java 类，编辑后的代码如下。

```
package cdtu.bean;
public class DateTimeDemo {
    private Integer id;
    private java.util.Date utilDate;
    private java.util.Date utilDate_timestamp;
    private java.util.Date utilDate_date;
    private java.util.Date utilDate_time;
    private java.sql.Date sqlDate;
    private java.sql.Time sqlTime;
    private java.sql.Timestamp sqlTimestamp;
    private java.util.Calendar utilCalendar;
    private java.util.Calendar utilCalendar_calendar;
    private java.util.Calendar utilCalendar_calendar_date;
    //省略 getters 和 setters
}
```

（2）使用 Hibernate Tools 在 cap.bean 子包中添加 DateTimeDemo.hbm.xml 映射文件。

```
<?xml version="1.0"?>
<!DOCTYPE hibernate-mapping PUBLIC "-//Hibernate/Hibernate Mapping DTD 3.0//EN"
"http://hibernate.sourceforge.net/hibernate-mapping-3.0.dtd">
<hibernate-mapping>
    <class name="cdtu.bean.DateTimeDemo" table="DATETIMEDEMO">
        <id name="id" type="java.lang.Integer">
            <column name="ID" />
            <generator class="native" />
        </id>
        <property name="utilDate" type="java.util.Date">
```

```
            <column name="UTILDATE" />
        </property>
        <property name="utilDate_timestamp" type="java.util.Date">
            <column name="UTILDATE_TIMESTAMP" />
        </property>
        <property name="utilDate_date" type="java.util.Date">
            <column name="UTILDATE_DATE" />
        </property>
        <property name="utilDate_time" type="java.util.Date">
            <column name="UTILDATE_TIME" />
        </property>
        <property name="sqlDate" type="java.sql.Date">
            <column name="SQLDATE" />
        </property>
        <property name="sqlTime" type="java.sql.Time">
            <column name="SQLTIME" />
        </property>
        <property name="sqlTimestamp" type="java.sql.Timestamp">
            <column name="SQLTIMESTAMP" />
        </property>
        <property name="utilCalendar" type="java.util.Calendar">
            <column name="UTILCALENDAR" />
        </property>
        <property name="utilCalendar_calendar" type="java.util.Calendar">
            <column name="UTILCALENDAR_CALENDAR" />
        </property>
        <property name="utilCalendar_calendar_date" type="java.util.Calendar">
            <column name="UTILCALENDAR_CALENDAR_DATE" />
        </property>
    </class>
</hibernate-mapping>
```

代码解释：<class>标签定义了 cdtu.bean 子包中 DateTimeDemo 持久化类对应数据库中的表为 DATETIMEDEMO，<id>标签对应了主键生成的方式，<property>对应了 DateTimeDemo 持久化类中属性与表 DATETIMEDEMO 中字段的对应关系。

（3）在 src 的 Hibernate 配置文件 hibernate.cfg.xml 中添加映射资源。

```
<mapping resource="cdtu/bean/DateTimeDemo.hbm.xml" />
```

（4）在 cap.test 中新建 DateTimeDemoTest.java 测试类，添加下面的测试方法。

```
@Test
    public void testDateTime() {
        Session sess = HibernateSessionFactory.getSession();
        Transaction tx = null;//第 1 行代码
        java.util.Date utilDate = new java.util.Date();
        java.sql.Date sqlDate = new java.sql.Date(0L);
        java.sql.Time sqlTime = new java.sql.Time(0L);
        java.sql.Timestamp sqlTimestamp = new java.sql.Timestamp(0L);
```

```
		java.util.Calendar utilCalendar = new java.util.GregorianCalendar();
		DateTimeDemo dateTimeTest = new DateTimeDemo();//第 7 行代码
		dateTimeTest.setUtilDate(utilDate);
		dateTimeTest.setUtilDate_timestamp(sqlTimestamp);
		dateTimeTest.setUtilDate_date(sqlDate);
		dateTimeTest.setUtilDate_time(sqlTime);
		dateTimeTest.setSqlDate(sqlDate);
		dateTimeTest.setSqlTime(sqlTime);
		dateTimeTest.setSqlTimestamp(sqlTimestamp);
		dateTimeTest.setUtilCalendar(utilCalendar);
		dateTimeTest.setUtilCalendar_calendar(utilCalendar);
		dateTimeTest.setUtilCalendar_calendar_date(utilCalendar); //第 17 行代码
		try {
			tx = sess.beginTransaction();//第 19 行代码
			sess.save(dateTimeTest);
			tx.commit();//第 21 行代码
		} catch (Exception e) {
			if (tx != null)
				tx.rollback();
			throw e;
		} finally {
			HibernateSessionFactory.closeSession();
		}
	}
```

代码解释：第 1 行代码定义 Transaction 对象 tx，第 2～6 行代码定义日期对象，第 7 行代码定义 DateTimeDemo 的对象 dateTimeTest，第 8～17 行代码调用 dateTimeTest 的 setters 方法设置属性值，第 19 行代码开启事务，第 20 行代码通过 session 将对象 dateTimeTest 转变为持久化对象。

5．和前一节相似，通过使用 Navicat，可以查看到生成的数据库表和添加的记录。

15.4 集合映射

15.4.1 Java 常用集合类

Java 的集合都包含在 java.util 包中，Java 集合中存放的是对象的引用，而非对象本身。Java 集合主要分为三种类型：

- Set：集合中的对象不按特定方式排序，并且没有重复对象，其实现类能对集合中的对象按特定方式排序。实现类有 HashSet、LinkedHashSet。
- List：集合中的对象按索引位置排序，可以有重复对象，允许按照对象在集合中的索引位置检索对象。实现类有 HashSet、LinkedHashSet。
- Map：集合中的每一个元素包含一对键/值对象，集合中没有重复的键对象，值对象可以重复。某些实现类能对集合中的键对象进行排序，实现类有 HashMap、Hashtable、

LinkedHashMap 等。

Java 集合类的具体使用可以参考相关 Java 文献，在这里就不提供具体的使用实例。

15.4.2 Hibernate 中集合元素

在 Hibernate 中集合映射的元素主要有：<list>元素用于映射 List 集合属性；<set>元素用于映射 Set 集合属性；<map>元素用于映射 Map 集合性；<bag>元素用于映射无序集合。

- <set>元素：可以映射类型为 java.util.Set 接口的属性，它的元素存放没有顺序且不允许重复；也可映射类型为 java.util.SortSet 接口的属性。它的元素可以按自然顺序排列。
- <list>元素：可以映射类型为 java.util.List 接口的属性。
- <bag>或<idbag>元素：可以映射类型为 java.util.Collection 接口的属性。它的元素可重复，但不保存顺序。
- <map>元素：可以映射类型为 java.util.Map 接口的属性，它的元素以键/值对的形式保存，也是无序的，也可以映射类型为 java.util.SortMap 接口的属性，它的元素可以按自然顺序排序。

下面将通过一个具体的实例讲解在 Hibernate 下集合的映射。

（1）继续在工程“hibernate3→src→cap.bean”下创建 CollectionAttributeDemo.java 类，编辑后的代码如下。

```
package cdtu.bean;
import java.util.ArrayList;
import java.util.HashMap;
import java.util.HashSet;
import java.util.List;
import java.util.Map;
import java.util.Set;
public class CollectionAttributeDemo {
    private Integer id;
    private String name;
    private List<String> schoolList = new ArrayList<String>();
    private Set<String> schoolSet = new HashSet<String>();
    private List<String> schoolBag = new ArrayList<String>();
    private Map<String, Integer> scoreMap = new HashMap<String, Integer>();
    @Override
    public String toString() {
        return "CollectionAttributeDemo [id=" + id + ", name=" + name
                + ", schoolList=" + schoolList + ", schoolSet=" + schoolSet
                + ", schoolBag=" + schoolBag + ", scoreMap=" + scoreMap + "]";
    }
    //省略 getters 和 setters

}
```

（2）使用 Hibernate Tools 在 cap.bean 子包中添加 CollectionAttributeDemo.hbm.xml 映射文件，编辑后的代码如下。

```
<?xml version="1.0"?>
<!DOCTYPE hibernate-mapping PUBLIC "-//Hibernate/Hibernate Mapping DTD 3.0//EN"
"http://hibernate.sourceforge.net/hibernate-mapping-3.0.dtd">
<hibernate-mapping>
    <class name="cdtu.bean.CollectionAttributeDemo" table="COLLECTION_ATTRIBUTE">
        <id name="id" type="java.lang.Integer">
            <column name="ID" />
            <generator class="native" />
        </id>
        <property name="name" type="java.lang.String">
            <column name="NAME" />
        </property>
        <list name="schoolList" inverse="false" table="SCHOOL_LIST" lazy="true">
            <key>
                <column name="ID" not-null="true" />
            </key>
            <list-index>
              <column name="LIST_ORDER" />
            </list-index>
            <element type="java.lang.String">
                <column name="SCHOOL_NAME" />
            </element>
        </list>
        <set name="schoolSet" table="SCHOOL_SET" inverse="false" lazy="true">
            <key>
                <column name="ID" not-null="true" />
            </key>
            <element type="java.lang.String">
                <column name="SCHOOL_NAME" not-null="true" />
            </element>
        </set>
        <bag name="schoolBag" table="SCHOOL_BAG" lazy="true">
            <key>
                <column name="ID" not-null="true" />
            </key>
            <element type="java.lang.String">
                <column name="SCHOOL_NAME" />
            </element>
        </bag>
        <map name="scoreMap" table="SCORE_MAP" lazy="true">
            <key>
                <column name="ID" not-null="true" />
            </key>
            <map-key type="java.lang.String">
              <column name="SUBJECT_NAME"></column>
            </map-key>
            <element type="java.lang.Integer">
                <column name="SCORE" />
```

```
            </element>
        </map>
    </class>
</hibernate-mapping>
```

代码解释：<list>标签将CollectionAttributeDemo类中的schoolList属性和表SCHOOL_LIST对应，表SCHOOL_LIST中会生成ID、SCHOOL_NAME和LIST_ORDER三个字段；<set>标签将CollectionAttributeDemo类中的schoolSet属性和表SCHOOL_SET对应，表SCHOOL_SET中会生成ID和SCHOOL_NAME两个字段，<bag>标签将CollectionAttributeDemo类中的schoolBag属性和表SCHOOL_BAG对应，表SCHOOL_BAG中会生成ID和SCHOOL_NAME两个字段，<map>标签将CollectionAttributeDemo类中的scoreMap属性和表SCORE_MAP对应，表SCORE_MAP中会生成ID、SCORE和SUBJECT_NAME三个字段，

（3）在src的Hibernate配置文件hibernate.cfg.xml中添加映射资源。

```
<mapping resource="cdtu/bean/CollectionAttributeDemo.hbm.xml" />
```

（4）在cap.test中新建EnumMappingTest.java测试类，添加下面的测试方法。

```
@Test
    public void testAdd() throws Exception {
        Session sess = HibernateSessionFactory.getSession();//第 1 行代码
        Transaction tx = null;
        CollectionAttributeDemo collectionAttr = new CollectionAttributeDemo();//第 3 行代码
        collectionAttr.setName("张三");
        collectionAttr.getSchoolList().add("成都七中");
        collectionAttr.getSchoolList().add("成都九中");
        collectionAttr.getSchoolSet().add("成都七中");
        collectionAttr.getSchoolSet().add("成都九中");
        collectionAttr.getSchoolBag().add("成都七中");
        collectionAttr.getSchoolBag().add("成都九中");
        collectionAttr.getScoreMap().put("语文", 140);
        collectionAttr.getScoreMap().put("数学", 145); //第 12 行代码
        try {
            tx = sess.beginTransaction();//第 14 行代码
            sess.save(collectionAttr);
            tx.commit();//第 16 行代码
        } catch (Exception e) {
            if (tx != null)
                tx.rollback();
            throw e;
        } finally {
            HibernateSessionFactory.closeSession();
        }
    }
```

代码解释：第1行代码通过工具类HibernateSessionFactory的静态方法获得Session对象sess，第3行代码创建CollectionAttributeDemo类的对象collectionAttr，第4～12行代码通过setters方法设置对象collectionAttr的属性值，第14行代码开启事务，第15行代码将对象

collectionAttr 转化为持久化对象，第 16 行代码提交事务。

15.5　实体对象关联关系映射

对象和关系数据是业务实体的两种表现形式，业务实体在内存中表现为对象，在数据库中表现为关系数据。因此在使用 ORM 的时候，不仅要映射具有不同类型属性的实体（数据库表），还要映射实体之间的关系（表关联）。

在面向对象的概念中，类与类之间的关系主要有泛化（Generalization）、依赖（Dependency）、关联（Association）、实现（Realization）四种。聚合（Aggregation）和组合（Composition）关系是关联关系的两种特例。关联关系是最普遍的关系，指两个类之间存在某种特定的对应关系，根据对应数量可以是一对一、一对多、多对多等。

关联关系是有方向性的，如果想表示每一张订单只属于一个客户的业务情境，可以建立起从 Order 到 Customer 的多对一关联，其方向为从“多”端（Order）指向“一”端（Customer），则每个 Order 对象都会引用一个 Customer 对象，因此在 Order 类中定义一个 Customer 类型的 customer 属性，来引用关联的 Customer 对象，如图 15-2 所示。

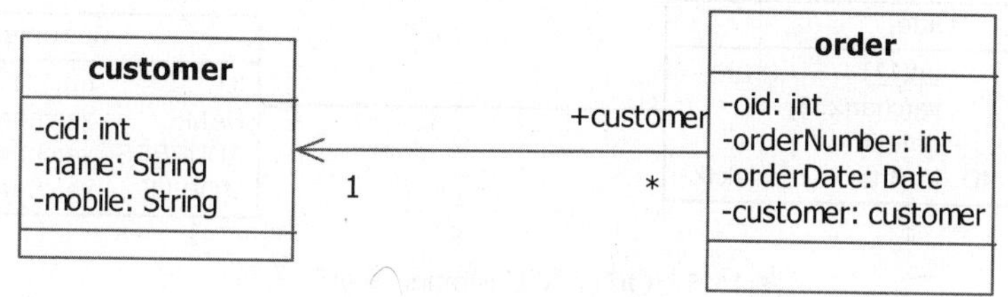

图 15-2　从 Order 到 Customer 的多对一单向关联

此外，如果想表示一个客户拥有多个订单的业务情境，也同样可以建立起从 Order 到 Customer 的多对一关联，但其方向变为从“一”端（Customer）指向“多”端（Order），则每个 Customer 对象都会引用多个 Order 对象，因此在 Customer 类中定义一个 Set 类型的 orders 属性，来引用关联的 Order 对象，如图 15-3 所示。

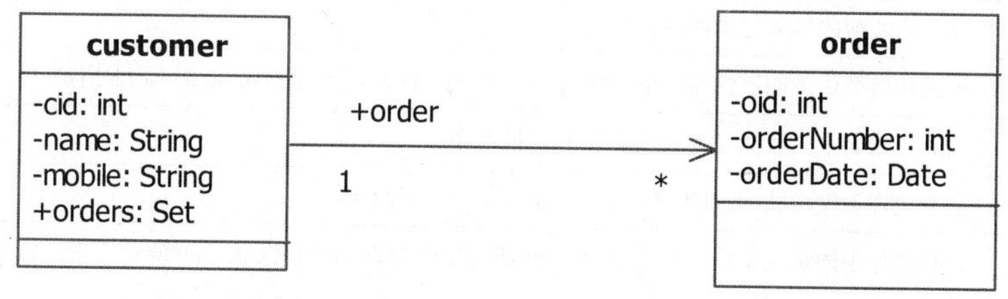

图 15-3　从 Customer 到 Order 的一对多单向关联

如果仅有从 Order 到 Customer 的关联，或者仅有从 Customer 到 Order 的关联，就称为单向关联。如果同时包含两种关联，则称为双向关联，如图 15-4 所示。

类间关系映射的通用规则应该使映射前后的多重性保持一致。因此一对多的对象关系映射成一对多的数据关系。

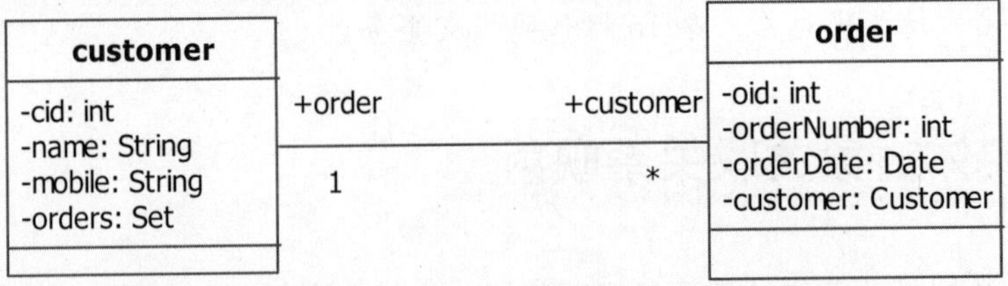

图 15-4　Customer 和 Order 的一对多双向关联

15.5.1　一对多关联关系映射

1. 关系数据模型

在关系数据库中，使用主外键参照关系表示表与表之间的关系，所有的关联都是双向的，不支持单向关联。以 Customer 和 Order 一对多关联为例，有了“one”端 Customer 的主键，可以根据“many”端 Order 表的外键查出“many”端数据；有了“many”端外键，可以根据“one”端表的主键查出“one”端数据，其中 CUSTOMER_ID 为 Customer 主键 ID 的别名，如图 15-5 所示。

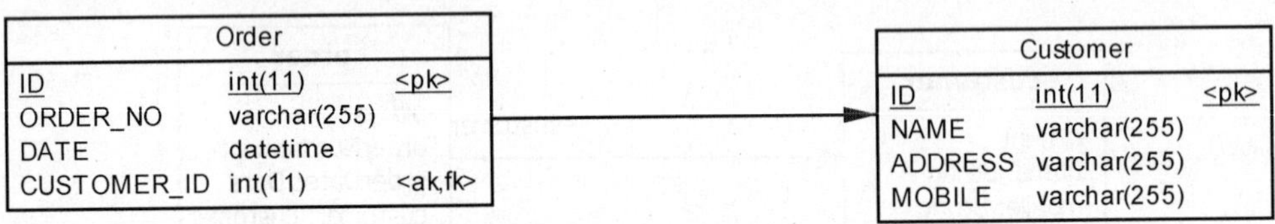

图 15-5　Order 和 Customer 多对一关系

Hibernate 使用<many-to-one>元素来映射多对一关联关系。由于使用< many-to-one >元素会把持久化类的属性映射为表的外键，所有<many-to-one >元素与<property>元素类似，它们的很多属性是相同的。

表 15-8　< many-to-one >元素的主要属性

属 性 名	功 能 说 明
name	设定待映射的持久化类的属性的名字，（必需）
column	设定和持久化类的属性对应的表的外键。它也可以通过嵌套的<column>元素指定
class	关联类的名字。默认是通过反射得到属性类型
not-null	是否允许为空。使用 DDL 为外键字段生成一个非空约束
cascade	设置操作中的级联策略。可选值为：all 所有操作情况均进行级联、none 所有操作情况均不进行级联、save-update 执行 save 和 update 操作时级联、delete 执行删除操作时级联
fetch	设置抓取数据的策略。默认值为：select 序列选择抓取（sequential select fetching）。可选值为：join 外连接抓取（outer-join fetching）
property-ref	指定关联类的一个属性，这个属性将会和本外键相对应（当外键参照一键时需要指定该属性）。如果没有指定，会使用对方关联类的主键

续表

属 性 名	功 能 说 明
unique	使用 DDL 为外键字段生成一个唯一约束。此外，这也可以用作 property-ref 的目标属性。这使关联同时具有一对一的效果
lazy	默认情况下，单点关联是经过代理的。lazy="true"，指定此属性在实例变量第一次被访问时应该延迟抓取（fetche lazily）（需要运行时字节码的增强）。lazy="false"指定此关联总是被预先抓取
not-found	指定外键引用的数据不存在时如何处理：ignore 会将数据不存在作为关联到一个空对象（null）处理。默认为 exception
access	Hibernate 用来访问属性的策略。默认是 property
optimistic-lock	指定这个属性在做更新时是否需要获得乐观锁定（optimistic lock）。换句话说，它决定这个属性发生脏数据时版本（version）的值是否增长
entity-name	被关联的类的实体名
insert/update	指定对应的字段是否包含在用于 UPDATE 和/或 INSERT 的 SQL 语句中。如果二者都是 false，则这是一个纯粹的“外源性（derived）”关联，它的值是通过映射到同一个（或多个）字段的某些其他属性得到，或者通过 trigger（触发器）、或其他程序

Hibernate 使用<one-to-many>元素映射一对多关联关系，建议在一端的 set 元素指定 inverse="true"，并且先插入一端对象，后插入多端对象，这样不会产生多余的 UPDATE 语句。

表 15-9 < one-to-many >元素的主要属性

属 性 名	功 能 说 明
class	关联类的名字。默认是通过反射得到属性类型
not-found	指定外键引用的数据不存在时如何处理：ignore 会将数据不存在作为关联到一个空对象（null）处理。默认为 exception
entity-name	被关联的类的实体名

2. 一对多双向映射实例

下面将通过具体的实例讲解 Hibernate 一对多双向映射的实现。

（1）打开 Eclipse 开发环境，选择菜单“File→New”，新建 Java Project，工程名为 Hibernate4，复制 HibernateSessionFactory 类到 cdap.util 包中，在 src 的 cdtu.bean.relation.one2many.bothway 新建 Cumstomer.java，编辑后的代码如下。

```
package cdtu.bean.relation.one2many.bothway;
import java.util.HashSet;
import java.util.Set;
public class Customer {
    private int id;
    private String name;
    private String address;
    private String mobile;
    private Set<Order> orders = new HashSet<Order>();//代码①
```

```
	public Customer() {
	}
	public Customer(String name, String address, String mobile) {
		this.name = name;
		this.address = address;
		this.mobile = mobile;
	}
	@Override
	public String toString() {
		return "Customer [id=" + id + ", name=" + name + ", address=" + address
				+ ", mobile=" + mobile + ", orders=" + orders + "]";
	}
//省略 getters 和 setters
}
```

代码解释：Customer 和 Order 是一对多关系，在 Customer 类的属性中使用 Set 结合表示。即代码①，表示一个 Customer 有多个 Order。

（2）在 src 的 cdtu.bean.relation.one2many.bothway 新建 Order.java，编辑后的代码如下。

```
package cdtu.bean.relation.one2many.bothway;
import java.util.Date;
public class Order {
	private int id;
	private String orderNo;
	private Date date;
	private Customer customer;//代码①
	public Order() {}
	public Order(String orderNo, Date date) {
		this.orderNo = orderNo;
		this.date = date;
	}
	@Override
	public String toString() {
		return "Order [id=" + id + ", orderNo=" + orderNo + ", date=" + date
				+ ", customer=" + customer + "]";
	}
// 省略 getters 和 setters
}
```

代码解释：Order 和 Customer 的关系是一对一，所以需要在 Order 类中添代码①，即一个 Order 只能对应一个客户。

（3）在 cdtu.bean.relation.one2many.bothway 对应包中使用 Hibernate Tools 生成映射文件 Cumstomer.hbm.xml，编辑后的代码如下。

```
<?xml version="1.0"?>
<!DOCTYPE hibernate-mapping PUBLIC "-//Hibernate/Hibernate Mapping DTD 3.0//EN"
"http://hibernate.sourceforge.net/hibernate-mapping-3.0.dtd">
<hibernate-mapping auto-import="false">
```

```
    <class                                   name="cdtu.bean.relation.one2many.bothway.Customer"
table="ONE2MANY_BOTHWAY_CUSTOMER">
        <id name="id" type="int">
            <column name="ID" />
            <generator class="native" />
        </id>
        <property name="name" type="java.lang.String">
            <column name="NAME" />
        </property>
        <property name="address" type="java.lang.String">
            <column name="ADDRESS" />
        </property>
        <property name="mobile" type="java.lang.String">
            <column name="MOBILE" />
        </property>
        <set name="orders" table="ORDER" inverse="true" lazy="true" order-by="DATE DESC">
            <key>
                <column name="CUSTOMER_ID" />
            </key>
            <one-to-many class="cdtu.bean.relation.one2many.bothway.Order" />
        </set>
    </class>
</hibernate-mapping>
```

代码解释：<set>标签映射 Customer 类的 orders 集合属性，对应的数据库表为 ORDER，属性 inverse=true，即由 Order 对象作为主控方，负责更新 order 表中的外键 CUSTOMER_ID 字段的值，lazy 属性为 true 表示实行懒加载，<one-to-many>标签指定一端的类为 Order。

（4）在 cdtu.bean.relation.one2many.bothway 对应包中使用 Hibernate Tools 生成映射文件 Order.hbm.xml，编辑后的代码如下。

```
<?xml version="1.0"?>
<!DOCTYPE hibernate-mapping PUBLIC "-//Hibernate/Hibernate Mapping DTD 3.0//EN"
"http://hibernate.sourceforge.net/hibernate-mapping-3.0.dtd">
<hibernate-mapping auto-import="false">
    <class name="cdtu.bean.relation.one2many.bothway.Order" table="ONE2MANY_BOTHWAY_ORDER">
        <id name="id" type="int">
            <column name="ID" />
            <generator class="native" />
        </id>
        <property name="orderNo" type="java.lang.String">
            <column name="ORDER_NO" />
        </property>
        <property name="date" type="java.util.Date">
            <column name="DATE" />
        </property>
        <many-to-one name="customer" class="cdtu.bean.relation.one2many.bothway.Customer" fetch="join">
            <column name="CUSTOMER_ID" />
        </many-to-one>
```

```
    </class>
</hibernate-mapping>
```

代码解释：<many-to-one>标签表示多对一的映射，class 属性指定对应一端的类为 Customer，外键为 CUSTOMER_ID.

（5）复制工程 hibernate3 工程中的配置文件 hibernate.cfg.xml 文件，将<mapping resource=…/>的内容修改为如下代码。

```
<!--映射 one2many-bothway 资源 -->
<mapping resource="cdtu/bean/relation/one2many/bothway/Customer.hbm.xml" />
<mapping resource="cdtu/bean/relation/one2many/bothway/Order.hbm.xml" />
```

（6）在 cdtu.test 中新建 JUnit 测试类 One2ManyBothWayTest.java，添加下面的测试用例。

```
@Test
    public void addTest() {
        Session session = HibernateSessionFactory.getSession();
        Transaction tx = null;
        Customer customer = new Customer("Benton", "1390000000", "四川成都");
        Order order1 = new Order("0001", new Date());
        Order order2 = new Order("0002", new Date());
        // 设定关联关系，代码①
        order1.setCustomer(customer);
        order2.setCustomer(customer);
        customer.getOrders().add(order1);
        customer.getOrders().add(order2);
//try catch 块为代码②
try {
            tx = session.beginTransaction();
            session.save(order1);
            session.save(order2);
            session.save(customer);
            tx.commit();
        } catch (Exception e) {
            if (tx != null)
                tx.rollback();
            throw e;
        } finally {
            HibernateSessionFactory.closeSession();
        }
    }
    @Test
    public void getTest() {
        Session session = HibernateSessionFactory.getSession();
        Transaction tx = null;
        try {
            tx = session.beginTransaction();
            Customer customer = (Customer) session.get(Customer.class, 1);
            System.out.println(customer.getName());
```

```
                System.out.println(customer.getOrders().getClass());
                System.out.println(customer.getOrders().size());
                tx.commit();
            } catch (Exception e) {
                if (tx != null)
                    tx.rollback();
                throw e;
            } finally {
                HibernateSessionFactory.closeSession();
            }
        }
        @Test
        public void updateTest() {
            Session session = HibernateSessionFactory.getSession();
            Transaction tx = null;
            try {
                tx = session.beginTransaction();
                    Order order = (Order) session.get(Order.class, 1);
order.getCustomer().setName("Tom");
                        tx.commit();
            } catch (Exception e) {
                if (tx != null)
                    tx.rollback();
                throw e;
            } finally {
                HibernateSessionFactory.closeSession();
            }
        }
        @Test
        public void deleteTest() {
            Session session = HibernateSessionFactory.getSession();
            Transaction tx = null;
            try {
                tx = session.beginTransaction();
                Customer customer = (Customer) session.get(Customer.class, 1);
                session.delete(customer);
                // 设定级联属性 cascade="delete"
                tx.commit();
            } catch (Exception e) {
                if (tx != null)
                    tx.rollback();
                throw e;
            } finally {
                HibernateSessionFactory.closeSession();
            }
        }
```

代码解释：addTest 方法中首先通过构造器创建一个 Customer 对象 customer，两个 Order

对象 order1 和 order2，通过代码①进行一对多的双向关联。代码②通过 session 的 save 方法将这些对象转变为持久化对象。

getTest 方法调用 session 的 get()方法查询 OID 为 1 的对象，然后打印 customer 对象的属性。

updateTest 方法，首先调用 session 的 get()方法查找 OID 为 1 的对象 order，然后通过 setter 方法修改 order 的属性，提交事务之后会更新触发数据库中的数据。

deleteTest 方法调用 session 的 delete()方法删除 OID 为 1 的对象，最终会实现删除数据库中主键为 1 的记录。

15.5.2　一对一关联关系映射

在 Hibernate 中，一对一关联关系的映射方式有两种：一种是基于外键的一对一关联关系映射；另一种是基于主键的一对一关联关系映射。接下来分别介绍这两种方式的实现。

1. 基于外键映射的一对一关联关系的映射

对于基于外键的一对一关联，其外键可以存放在任意一边。在需要外键的一端，增加<many-to-one>元素，并为<many-to-one>元素增加 unique="true"属性来表示为一对一关联关系。在无外键的一端需要使用<one-to-one>元素，该元素使用 property-ref 属性指定关联类的属性名，此属性和本类的主键相对应。<one-to-one>的主要属性如表 15-10 所示。

表 15-10　<one-to-one>元素的主要属性

属 性 名	功 能 说 明
name	必须，设定待映射的持久化类的属性的名字
column	设定和持久化类的属性对应的表的外键
property-ref	设置关联类的属性名，此属性和本类的主键相对应。默认值为关联类的主键
cascade	设置操作中的级联策略。可选值为：all 所有操作情况均进行级联、none 所有操作情况均不进行级联、save-update 执行更新操作时级联、delete 执行删除操作时级联
constrained	表明当前类对应的表与被关联的表之间是否存在着外键约束。默认值为 false
fetch	设置抓取数据的策略。可选值为：join 外连接抓取、select 序列选择抓取
not-null	是否允许为空
lazy	指定是否采用延迟加载及加载策略。默认值为 proxy，通过代理进行关联。可选值为 true，此对象采用延迟加载并在变量第一次被访问时抓取；值为 false 表示此关联对象不采用延迟加载
access	Hibernate 访问这个属性的策略。默认值为 property
entity-name	被关联类的实体名
formula	绝大多数一对一关联都指向其实体的主键。在某些情况下会指向一个或多个字段或是一个表达式，此时可用一个 SQL 公式来表示

关系数据模型

一对一的对象关系映射成一对一的数据关系。假设客户（Customer）和客户联系地址（Address）是一对一的关系，每个客户仅有一个客户联系地址，每个客户联系地址对应唯一的客户，传统的数据关系如图 15-6 所示。

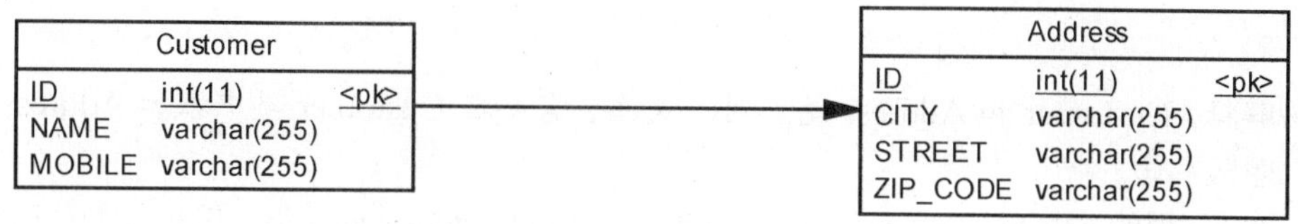

图 15-6 Customer 和 Address 一对一关系

有了“one”端 Address 的主键 ID，可以根据另一个“one”端 Customer 的外键查出 Customer 的数据，反向同理。

因为一对一的关系是一对多关系的子集，一对多关系也是多对多关系的子集，因此也可以将一对一的对象关系映射成一对多的关系或者多对多的关系。

Hibernate 基于外键映射的一对一关联关系之后的关系数据模型如图 15-7 所示，其中 ADDERSS_ID 为 Address 表主键 ID 的别名。

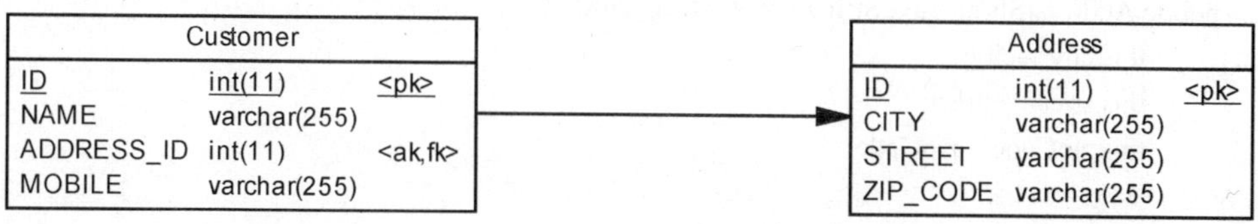

图 15-7 基于外键映射的一对一关联关系的数据模型

Customer 端的外键需要建立唯一性约束，从而保证一对一关系。因为映射过程可以将外键放在 Customer 一边，也可以放在 Address 一边，该图展示的是将外键放在 Customer 里。

基于外键的一对一双向映射实例

下面将通过具体的实例讲解基于外键的一对一双向映射实例。

（1）继续使用工程 hibernate4，在 src 中的 cdtu.bean.relation.one2one.fk 子包新建 Customer.java 类，编辑后的代码如下。

```
package cdtu.bean.relation.one2one.fk;
public class Customer {
    private Integer id;
    private String name;
    private Address address;//代码①
    private String mobile;
    public Customer() {
    }
    public Customer(String name, String mobile) {
        this.name = name;
        this.mobile = mobile;
    }
    @Override
    public String toString() {
        return "Customer [id=" + id + ", name=" + name + ", address=" + address
                + ", mobile=" + mobile + "]";
    }
// 省略 getters 和 setters
```

```
}
```

代码解释，Customer 和 Address 是一对一关系，需要在 Customer 类中添加 Address 的对象 address，即代码①。

（2）在 src 中的 cdtu.bean.relation.one2one.fk 子包新建 Address.java 类，编辑后的代码如下。

```
package cdtu.bean.relation.one2one.fk;
public class Address {
        private Integer id;
        private String city;
        private String street;
        private String zipCode;
        private Customer customer;//代码①
        public Address() {}
        public Address(String city, String street, String zipCode) {
                this.city = city;
                this.street = street;
                this.zipCode = zipCode;
        }
        @Override
        public String toString() {
                return "Address [id=" + id + ", city=" + city + ", street=" + street
                                + ", zipCode=" + zipCode + "]";
        }
//省略 getters 和 setters
}
```

代码解释：Address 和 Customer 也是一对一关系，需要在 Address 类中添加 Customer 的对象 customer，即代码①。

（3）在 cdtu.bean.relation.one2one.fk 对应包中使用 Hibernate Tools 生成映射文件 Cumstomer.hbm.xml，编辑后的代码如下。

```
<?xml version="1.0"?>
<!DOCTYPE hibernate-mapping PUBLIC "-//Hibernate/Hibernate Mapping DTD 3.0//EN"
"http://hibernate.sourceforge.net/hibernate-mapping-3.0.dtd">
<hibernate-mapping auto-import="false">
        <class name="cdtu.bean.relation.one2one.fk.Customer" table="One2One_FK_CUSTOMER">
                <id name="id" type="java.lang.Integer">
                        <column name="ID" />
                        <generator class="native" />
                </id>
                <property name="name" type="java.lang.String">
                        <column name="NAME" />
                </property>
                <many-to-one name="address" class="cdtu.bean.relation.one2one.fk.Address"
                        cascade="all" fetch="join">
                        <column name="ADDRESS_ID" unique="true" />
```

```
            </many-to-one>
            <property name="mobile" type="java.lang.String">
                <column name="MOBILE" />
            </property>
        </class>
    </hibernate-mapping>
```

代码解释：<many-to-one>标签指定外键所处的一端，属性为 address，对应的类为 Address，使用 unique="true"属性来表示为一对一关联关系。

（4）在 cdtu.bean.relation.one2one.fk 对应包中使用 Hibernate Tools 生成映射文件 Address.hbm.xml，编辑后的代码如下。

```
    <?xml version="1.0"?>
    <!DOCTYPE hibernate-mapping PUBLIC "-//Hibernate/Hibernate Mapping DTD 3.0//EN"
    "http://hibernate.sourceforge.net/hibernate-mapping-3.0.dtd">
    <hibernate-mapping auto-import="false">
        <class name="cdtu.bean.relation.one2one.fk.Address" table="One2One_FK_ADDRESS">
            <id name="id" type="java.lang.Integer">
                <column name="ID" />
                <generator class="native" />
            </id>
            <property name="city" type="java.lang.String">
                <column name="CITY" />
            </property>
            <property name="street" type="java.lang.String">
                <column name="STREET" />
            </property>
            <property name="zipCode" type="java.lang.String">
                <column name="ZIP_CODE" />
            </property>
            <one-to-one   name="customer"   class="cdtu.bean.relation.one2one.fk.Customer"   property-ref=
"address">
    </one-to-one>
        </class>
    </hibernate-mapping>
```

代码解释：<one-to-one>使用在无外键的一端，属性为 customer，对应的类为 Customer，使用 property-ref 属性指定关联类的属性名为 address。

（5）在 src 的根目录下的 Hibernate 配置文件 hibernate.cfg.xml 中，添加下面的映射资源。

```
    <!--映射 one2one-fk 资源 -->
    <mapping resource="cdtu/bean/relation/one2one/fk/Customer.hbm.xml" />
    <mapping resource="cdtu/bean/relation/one2one/fk/Address.hbm.xml" />
```

（6）在 cdtu.test 子包中新建 Junit 测试类 One2OneFkTest.java 类，添加下面的测试方法。

```
    @Test
        public void testAdd()
        {
```

```
        Session session = HibernateSessionFactory.getSession();
        Transaction tx = null;
        //代码①
        Customer customer = new Customer("Benton","1390000000");
        Address address = new Address("成都","郫县","610000");
        customer.setAddress(address);
        address.setCustomer(customer);
        //代码②
        try {
            tx = session.beginTransaction();
            session.save(customer);
            tx.commit();
        } catch (Exception e) {
            if(tx!=null) tx.rollback();
            throw e;
        } finally {
            HibernateSessionFactory.closeSession();
        }
    }
    @Test
    public void testGet()
    {
        Session session = HibernateSessionFactory.getSession();
        Transaction tx = null;
        try {
            tx = session.beginTransaction();
            //1. 默认情况下对关联属性使用懒加载
            Customer customer = (Customer)session.get(Customer.class, 1);
            System.out.println(customer);
            //2. 对关联属性使用懒加载时，可能出现懒加载异常的问题
            HibernateSessionFactory.closeSession();
            Address address = customer.getAddress();
            System.out.println(address.getClass());
            System.out.println(address.getCity());
            //Address address = (Address)session.get(Address.class, 1);
            //System.out.println(address);
            tx.commit();
        } catch (Exception e) {
            if(tx!=null) tx.rollback();
            throw e;
        } finally {
            HibernateSessionFactory.closeSession();
        }
    }
```

代码解释：在 testAdd 方法中，代码①首先创建两个对象 customer 和 address，接着设置 customer 对象和 address 对象一对一的双向关联关系。代码②调用 session 的 save 方法将 customer 对象转化为持久化对象。

testGet()方法首先调用 session 的 get()方法查找 OID 为 1 的 Customer 对象 customer，然后通过对象导航查询 customer 对象对应的 address 对象信息，并且打印出 address 对象的相关信息。

2. 基于主键映射的一对一关联关系的映射

基于主键的一对一关联关系的映射策略是：一端的主键生成器使用 foreign 策略，表明根据“对方”的主键来生成自己的主键，自己并不能独立生成主键。使用<param>子元素指定使用当前持久化类的哪个属性作为“对方”。

采用 foreign 主键生成器策略的一端增加<one-to-one>元素映射关联属性，其<one-to-one>属性还应增加 constrained="true"属性。

constrained（约束）：指定为当前持久化类对应的数据库表的主键添加一个外键约束，引用被关联的对象（“对方”）所对应的数据库表主键。

另一端增加 one-to-one 元素映射关联属性，并设置相应的 cascade 属性值。

关系数据模型

一对一的对象关系映射成一对一的数据关系。假设客户（Customer）和客户联系地址（Address）是一对一的关系，每个客户仅有一个客户联系地址，每个客户联系地址对应唯一的客户。基于主键映射的一对一关联关系的数据模型如图 15-8 所示。Address 根据 Customer 端的主键来生成自己的主键，需要建立 Address 端 ID 的非空和唯一性约束，以保证一对一关系。

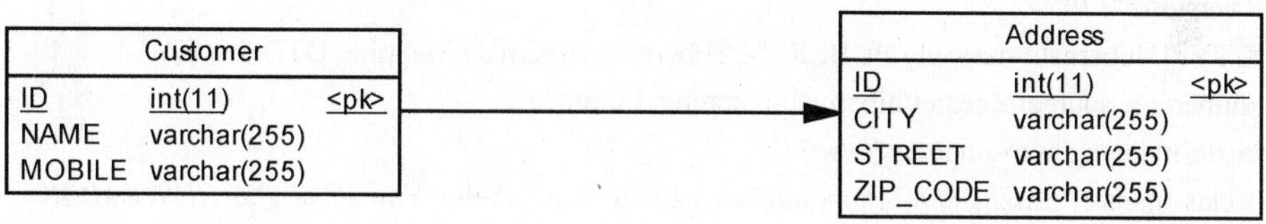

图 15-8　基于主键映射的一对一关联关系的数据模型

基于主键的一对一双向映射案例

下面将通过具体的实例讲解基于主键的一对一双向映射。

（1）复制 cdtu.bean.relation.one2one.fk 下的 Customer.java，Address.java，以及 Customer.hbm.xml，Address.hbm.xm 到 cdtu.bean.relation.one2one.pk 子包中。

（2）修改 Address.hbm.xm 映射文件，代码如下。

```
<?xml version="1.0"?>
<!DOCTYPE hibernate-mapping PUBLIC "-//Hibernate/Hibernate Mapping DTD 3.0//EN"
"http://hibernate.sourceforge.net/hibernate-mapping-3.0.dtd">
<hibernate-mapping auto-import="false">
    <class name="cdtu.bean.relation.one2one.pk.Address" table="One2One_PK_ADDRESS">
        <id name="id" type="java.lang.Integer">
            <column name="ID" />
            <generator class="foreign">
                <param name="property">customer</param>
            </generator>
        </id>
        <property name="city" type="java.lang.String">
            <column name="CITY" />
        </property>
```

```
            <property name="street" type="java.lang.String">
                <column name="STREET" />
            </property>
            <property name="zipCode" type="java.lang.String">
                <column name="ZIP_CODE" />
            </property>
            <one-to-one name="customer" class="cdtu.bean.relation.one2one.pk.Customer"
                constrained="true">
    </one-to-one>
        </class>
    </hibernate-mapping>
```

代码解释：<id>标签中的子标签<generator>属性 class="foreign"，表明根据对方 customer 的主键来生成自己的主键，自己并不能独立生成主键。使用<param>子元素指定使用当前持久化类的 customer 属性作为对方。

同时在<one-to-one>标签中指定属性为 customer，对应的类为 Customer，并设置 constrained="true"属性。

（3）修改 Customer.hbm.xml 映射文件，代码如下。

```
<?xml version="1.0"?>
<!DOCTYPE hibernate-mapping PUBLIC "-//Hibernate/Hibernate Mapping DTD 3.0//EN"
"http://hibernate.sourceforge.net/hibernate-mapping-3.0.dtd">
<hibernate-mapping auto-import="false">
    <class name="cdtu.bean.relation.one2one.pk.Customer" table="One2One_PK_CUSTOMER">
        <id name="id" type="java.lang.Integer">
            <column name="ID" />
            <generator class="native" />
        </id>
        <property name="name" type="java.lang.String">
            <column name="NAME" />
        </property>
        <one-to-one name="address" class="cdtu.bean.relation.one2one.pk.Address"
            cascade="all"></one-to-one>
        <property name="mobile" type="java.lang.String">
            <column name="MOBILE" />
        </property>
    </class>
</hibernate-mapping>
```

代码解释：<one-to-one>标签中属性为 address 对应 Customer 中的对象 address，对应的类为 Address。

（4）在 src 根目录下的 Hibernate 配置文件 hibernate.cfg.xml 中，添加下面的映射资源。

```
<!--映射 one2one-pk 资源 -->
<mapping resource="cdtu/bean/relation/one2one/pk/Customer.hbm.xml" />
<mapping resource="cdtu/bean/relation/one2one/pk/Address.hbm.xml" />
```

（5）在 cdtu.cap 中新建 One2OnePkTest 测试类，添加下面的测试方法。

```
@Test
    public void testAdd() {
        Session session = HibernateSessionFactory.getSession();
        Transaction tx = null;
        //代码块①
        Customer customer = new Customer("Benton", "1390000000");
        Address address = new Address("成都", "郫县", "610000");
        customer.setAddress(address);
        address.setCustomer(customer);
        //代码块②
        try {
            tx = session.beginTransaction();
        // 注意：可以保存 address、customer 或两个都保存，都能正确保存（都会生成两条 SQL 语句）
        // session.save(address);
            session.save(customer);
            tx.commit();
        } catch (Exception e) {
            if (tx != null)
                tx.rollback();
            throw e;
        } finally {
            HibernateSessionFactory.closeSession();
        }
    }
    @Test
    public void testGet()
    {
        Session session = HibernateSessionFactory.getSession();
        Transaction tx = null;
        try {
            tx = session.beginTransaction();
            Customer customer = (Customer)session.get(Customer.class, 1);
            System.out.println(customer);
            //Address address = (Address)session.get(Address.class, 1);
            //System.out.println(address);
            tx.commit();
        } catch (Exception e) {
            if(tx!=null) tx.rollback();
            throw e;
        } finally {
            HibernateSessionFactory.closeSession();
        }
    }
```

代码解释：testAdd()方法中，代码块①通过构造器创建 customer 和 address 对象，接着设置一对一的双向关联关系，代码块②调用 session 的 save()方法将 customer 对象转变为持久化对象。

testGet()方法调用 session 的 get()方法查找 Customer 对象中 OID 为 1 的对象，然后打印其信息。

15.5.3 多对多关联关系映射

在实际开发中，实现数据库表之间的多对多关联关系有两种方法。一种是传统的多对多映射，另外一种是将多对多关系拆分为两个一对多关联关系。下面分别讲解双向多对多关联关系映射和使用两个一对多关联关系映射代替一个多对多关系映射。

Hibernate 中映射多对多关系要使用到< many-to-many>元素，< many-to-many>元素的主要属性如表 15-11 所示。

表 15-11 < many-to-many>元素的主要属性

属 性 名	功 能 说 明
column	关联的字段。设定和持久化类的属性对应的表的外键。它也可以通过嵌套的<column>元素指定
class	关联类的名字。默认是通过反射得到属性类型
fetch	设置抓取数据的策略。默认值为 select 序列选择抓取。可选值为 join 外连接抓取。在外连接抓取（outer-join fetching）和序列选择抓取（sequential select fetching）两者中选择其一
property-ref	指定关联类的一个属性，这个属性将会和本外键相对应（当外键参照唯一键时需要指定该属性）。如果没有指定，会使用对方关联类的主键
unique	使用 DDL 为外键字段生成一个唯一约束。此外，这也可以用作 property-ref 的目标属性。这使关联同时具有一对一的效果
lazy	默认情况下，单点关联是经过代理的。lazy="true"，指定此属性应该在实例变量第一次被访问时进行延迟抓取（fetche lazily）（需要运行时字节码的增强）。lazy="false"，指定此关联总是被预先抓取
not-found	指定外键引用的数据不存在时如何处理：ignore 会将数据不存在作为关联到一个空对象（null）处理。默认为 exception
entity-name	被关联的类的实体名

多对多的关联必须使用连接表，两个表的主键构成连接表的联合主键。双向多对多关联需要两端都使用集合属性。集合属性应增加 key 子元素用以映射外键列，集合元素里还应增加<many-to-many>子元素关联实体类。

对于双向多对多关联，必须把其中一端的 inverse 属性设置为 true，否则两端都维护关联关系可能会造成主键冲突。

传统多对多映射

关系数据模型

关系数据库中的多对多关系可以通过两个一对多关系呈现。假设订单（Order）和产品（Product）之间是多对多的关系，每笔订单订购多种产品，每种产品对应多个订单，其关系如图 15-9 所示。

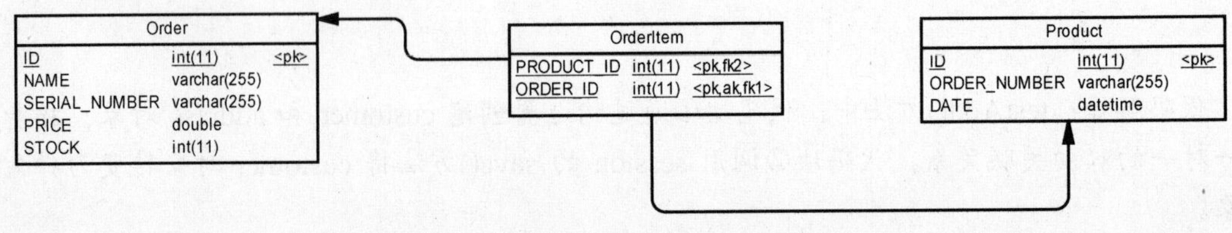

图 15-9 Order 和 Product 多对多关系

Hibernate 传统的多对多映射之后的数据模型也如图 15-9 所示类似，OrderItem 是多对多关联使用的连接表，两个表的主键构成连接表的联合主键。

多对多双向映射实例

下面将通过具体的实例讲解双向多对多的关系映射。

（1）继续使用工程 hibernate4，在 src 中的 cdtu.bean.relation.many2many.bothway 子包新建 Order.java 类，编辑后的代码如下。

```
package cdtu.bean.relation.many2many.bothway;
import java.util.Date;
import java.util.HashSet;
import java.util.Set;
public class Order {
        private Integer id;// 订单 id
        private String orderNumber;// 订单号
        private Date date;// 下单时间
        private Set<Product> products = new HashSet<Product>();// 产品集合，代码①
        public Order() {}
        public Order(String orderNumber, Date date) {
                this.orderNumber = orderNumber;
                this.date = date;
        }
        @Override
        public String toString() {
                return "Order [id=" + id + ", orderNumber=" + orderNumber + ", date="
                                + date + ", products=" + products + "]";
        }
//省略 getters 和 setters
}
```

（2）在 src 中的 cdtu.bean.relation.many2many.bothway 子包新建 Product.java 类，编辑后的代码如下。

```
package cdtu.bean.relation.many2many.bothway;
import java.util.HashSet;
import java.util.Set;
public class Product {
        private Integer id;// 产品 id
        private String name;// 产品名称
        private String serialNumber;// 产品序列号
        private double price;// 产品价格
        private int stock;// 产品库存量
        private Set<Order> orders = new HashSet<Order>();// 订单项代码①

        public Product() {}
        public Product(String name, String serialNumber, double price, int stock) {
                this.name = name;
                this.serialNumber = serialNumber;
                this.price = price;
```

```
            this.stock = stock;
        }
        @Override
        public String toString() {
            return "Product [id=" + id + ", name=" + name + ", serialNumber="
                    + serialNumber + ", price=" + price + ", stock=" + stock
                    + ", orders=" + orders + "]";
        }
        //省略 getters 和 setters
}
```

在 cdtu.bean.relation.many2many.bothway 对应的包中使用 Hibernate Tools 生成映射文件 Order.hbm.xml，编辑后的代码如下。

```
<?xml version="1.0"?>
<!DOCTYPE hibernate-mapping PUBLIC "-//Hibernate/Hibernate Mapping DTD 3.0//EN"
"http://hibernate.sourceforge.net/hibernate-mapping-3.0.dtd">
<hibernate-mapping package="cdtu.bean.relation.many2many.bothway" auto-import="false">
    <class name="Order" table="N2N_BOTHWAY_ORDER">
        <id name="id" type="java.lang.Integer">
            <column name="ID" />
            <generator class="native" />
        </id>
        <property name="orderNumber" type="java.lang.String">
            <column name="ORDER_NUMBER" />
        </property>
        <property name="date" type="java.util.Date">
            <column name="DATE" />
        </property>
        <set name="products" table="N2N_BOTHWAY_ORDER_PRODUCT" inverse="true" lazy="true">
            <key>
                <column name="ORDER_ID" />
            </key>
            <many-to-many class="Product">
              <column name="PRODUCT_ID" ></column>
            </many-to-many>
        </set>
    </class>
</hibernate-mapping>
```

代码解释：<set>标签中的 name 属性 products 对应于 Order 类中的代码①，中间表为 N2N_BOTHWAY_ORDER_PRODUCT，inverse="true"表示 Order 为主控方，lazy="true"表示是懒加载，<key>标签中的<column>指定中间表 N2N_BOTHWAY_ORDER_PRODUCT 的外键字段为 ORDER_ID，<many-to-many>标签指定 Order 对应的多方类为 Product，<column>标签指定中间表 N2N_BOTHWAY_ORDER_PRODUCT 的外键字段为 PRODUCT_ID。

（4）在 cdtu.bean.relation.many2many.bothway 对应包中使用 Hibernate Tools 生成映射文件 Product.hbm.xml，编辑后的代码如下。

```
<?xml version="1.0"?>
<!DOCTYPE hibernate-mapping PUBLIC "-//Hibernate/Hibernate Mapping DTD 3.0//EN"
"http://hibernate.sourceforge.net/hibernate-mapping-3.0.dtd">
<hibernate-mapping package="cdtu.bean.relation.many2many.bothway" auto-import="false">
    <class name="Product" table="N2N_BOTHWAY_PRODUCT">
        <id name="id" type="java.lang.Integer">
            <column name="ID" />
            <generator class="native" />
        </id>
        <property name="name" type="java.lang.String">
            <column name="NAME" />
        </property>
        <property name="serialNumber" type="java.lang.String">
            <column name="SERIAL_NUMBER" />
        </property>
        <property name="price" type="double">
            <column name="PRICE" />
        </property>
        <property name="stock" type="int">
            <column name="STOCK" />
        </property>
        <set name="orders" table="N2N_BOTHWAY_ORDER_PRODUCT" inverse="false"
            lazy="true">
            <key>
                <column name="PRODUCT_ID" />
            </key>
            <many-to-many class="Order">
                <column name="ORDER_ID" />
            </many-to-many>
        </set>
    </class>
</hibernate-mapping>
```

代码解释：<set>标签中的 name="orders"对应于 Product 类中的代码①，中间表为 N2N_BOTHWAY_ORDER_PRODUCT，inverse="false"表示 Order 为主控方，lazy="true"表示是懒加载，<key>标签中的<column>指定中间表 N2N_BOTHWAY_ORDER_PRODUCT 的外键字段为 PRODUCT_ID，<many-to-many>标签指定 Product 对应的多方类为 Order，<column>标签指定中间表 N2N_BOTHWAY_ORDER_PRODUCT 的外键字段为 ORDER_ID。

（5）在 src 的根目录下的 Hibernate 配置文件 hibernate.cfg.xml 中，添加下面的映射资源。

```
<!--映射 many2many-bothway 资源 -->
<mapping resource="cdtu/bean/relation/many2many/bothway/Product.hbm.xml" />
<mapping resource="cdtu/bean/relation/many2many/bothway/Order.hbm.xml" />
```

在 cdtu.test 子包中新建 Junit 测试类 BothwayN2NTest.java 类，添加下面的测试方法。

```
@Test
    public void testAdd() {
```

```
        Session session = HibernateSessionFactory.getSession();
        Transaction tx = null;
        //代码块①
        Order order1 = new Order("00001", new Date());
        Order order2 = new Order("00002", new Date());
        Product product1 = new Product("Java Web 技术", "100001", 60, 100);
        Product product2 = new Product("Java EE 技术", "100001", 80, 100);
        // 设置关联关系
        order1.getProducts().add(product1);
        order1.getProducts().add(product2);
        order2.getProducts().add(product1);
        product1.getOrders().add(order1);
        product2.getOrders().add(order1);
        product1.getOrders().add(order2);
        //代码块②
        try {
            tx = session.beginTransaction();
            session.save(order1);
            session.save(order2);
            session.save(product1);
            session.save(product2);
            tx.commit();
        } catch (Exception e) {
            if (tx != null)
                tx.rollback();
            throw e;
        } finally {
            HibernateSessionFactory.closeSession();
        }
    }
```

代码解释：代码块①通过各自的构造器创建对象 product1、product2、order1，和 order2 对象，余下的代码设置多对多的双向关联关系。代码块②调用 session 的 save()方法将这些对象转变为持久化对象。

两个一对多实现多对多映射

关系数据模型

拆分多对多关联关系为两个一对多关联关系，可以使用另一种方式，即关联关系中的 orderItem 映射成独立的表，具有独立的主键和相关属性，而不是传统多对多映射中的关联表。其数据模型如图 15-10 所示。

基于两个一对多实现多对多双向映射的实例

下面将通过具体的实例讲解用两个一对多关系实现多对多的映射，实例数据模型如图 15-10 所示，由于一对多的双向关联关系在 15.5.1 节（一对多关联关系映射）已经讲过，所以不再提供代码解释。

（1）继续使用工程 hibernate4，在 src 中的 cdtu.bean.relation.many2many.doubleone2many 子包新建 Customer.java 类，编辑后的代码如下。

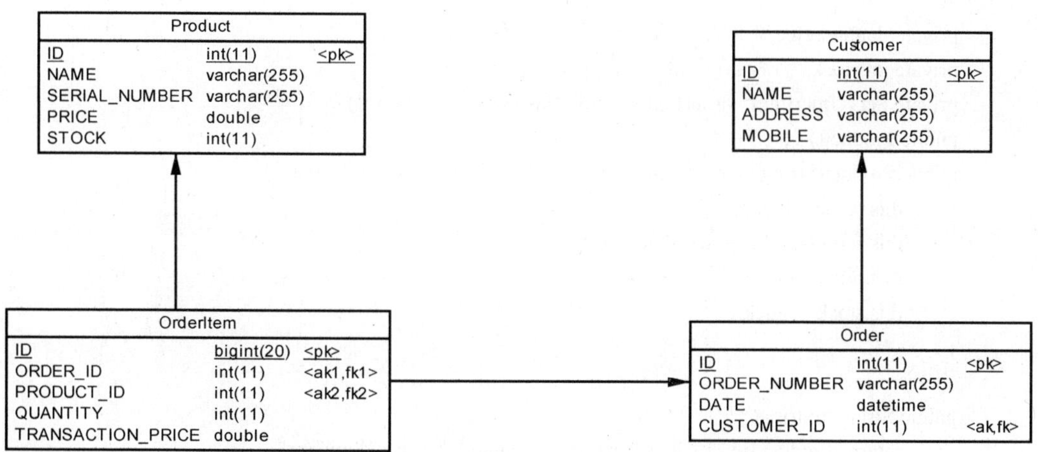

图 15-10 拆分多对多关联关系为两个一对多关联关系数据模型

```
package cdtu.bean.relation.many2many.doubleone2many;
import java.util.HashSet;
import java.util.Set;
public class Customer {
    private int id;//客户 id
    private String name;//姓名
    private String address;//地址
    private String mobile;//手机号码
    private Set<Order> orders = new HashSet<Order>();//订单
    public Customer() {}
    public Customer(String name, String address, String mobile) {
        this.name = name;
        this.address = address;
        this.mobile = mobile;
    }
    @Override
    public String toString() {
        return "Customer [id=" + id + ", name=" + name + ", address=" + address
                + ", mobile=" + mobile + ", orders=" + orders + "]";
    }
//省略 getters 和 setters
}
```

（2）在 src 的 cdtu.bean.relation.many2many.doubleone2many 子包中新建 Product.java 类，编辑后的代码如下。

```
package cdtu.bean.relation.many2many.doubleone2many;
import java.util.HashSet;
import java.util.Set;
public class Product {
    private Integer id;// 产品 id
    private String name;// 产品名称
    private String serialNumber;// 产品序列号
```

```
    private double price;// 产品价格
    private int stock;// 产品库存量
    private Set<OrderItem> orderItems = new HashSet<OrderItem>();// 订单项
    public Product() {}
    public Product(String name, String serialNumber, double price, int stock) {
        this.name = name;
        this.serialNumber = serialNumber;
        this.price = price;
        this.stock = stock;
    }
    @Override
    public String toString() {
        return "Product [id=" + id + ", name=" + name + ", serialNumber="
                + serialNumber + ", price=" + price + ", stock=" + stock
                + ", orderItems=" + orderItems + "]";
    }
//省略 getters 和 setters
}
```

（3）在 src 的 cdtu.bean.relation.many2many.doubleone2many 子包中新建 Order.java 类，编辑后的代码如下。

```
package cdtu.bean.relation.many2many.doubleone2many;
import java.util.Date;
import java.util.HashSet;
import java.util.Set;
public class Order {
    private Integer id;// 订单 id
    private String orderNumber;// 订单号
    private Date date;// 下单时间
    private Customer customer;// 客户
    private Set<OrderItem> orderItems = new HashSet<OrderItem>();// 订单项
    public Order() {}
    public Order(String orderNumber, Date date) {
        this.orderNumber = orderNumber;
        this.date = date;
    }
    @Override
    public String toString() {
        return "Order [id=" + id + ", orderNumber=" + orderNumber + ", date="
                + date + ", customer=" + customer + ", orderItems="
                + orderItems + "]";
    }
    //省略 getters 和 setters
}
```

（4）在 src 的 cdtu.bean.relation.many2many.doubleone2many 子包中新建 OrderItem.java 类，编辑后的代码如下。

```
package cdtu.bean.relation.many2many.doubleone2many;
public class OrderItem {
    private Long id;// 订单项 id
    private Order order;// 订单
    private Product product;// 产品
    private int quantity;// 数量
    private double transactionPrice;// 成交价
    public OrderItem() {}
    public OrderItem(Order order, Product product, int quantity) {
        this.order = order;
        this.product = product;
        this.quantity = quantity;
    }
    @Override
    public String toString() {
        return "OrderItem [id=" + id + ", order=" + order + ", product="
                + product + ", quantity=" + quantity + ", transactionPrice="
                + transactionPrice + "]";
    }
// 省略 getters 和 setters
}
```

（5）在 cdtu.bean.relation.many2many.doubleone2many 对应包中使用 Hibernate Tools 生成映射文件 Cumstomer.hbm.xml，编辑后的代码如下。

```
<?xml version="1.0"?>
<!DOCTYPE hibernate-mapping PUBLIC "-//Hibernate/Hibernate Mapping DTD 3.0//EN"
"http://hibernate.sourceforge.net/hibernate-mapping-3.0.dtd">
<hibernate-mapping auto-import="false">
    <class name="cdtu.bean.relation.many2many.doubleone2many.Customer"
     table="N2N_DOUBLE_1_N_CUSTOMER">
        <id name="id" type="int">
            <column name="ID" />
            <generator class="native" />
        </id>
        <property name="name" type="java.lang.String">
            <column name="NAME" />
        </property>
        <property name="address" type="java.lang.String">
            <column name="ADDRESS" />
        </property>
        <property name="mobile" type="java.lang.String">
            <column name="MOBILE" />
        </property>
        <set name="orders" table="DOUBLE_1_N_ORDER" inverse="true" lazy="true">
            <key>
                <column name="CUSTOMER_ID" />
            </key>
```

```
                <one-to-many class="cdtu.bean.relation.many2many.doubleone2many.Order" />
            </set>
        </class>
</hibernate-mapping>
```

（6）在 cdtu.bean.relation.many2many.doubleone2many 对应包中使用 Hibernate Tools 生成映射文件 Product.hbm.xml，编辑后的代码如下。

```
<?xml version="1.0"?>
<!DOCTYPE hibernate-mapping PUBLIC "-//Hibernate/Hibernate Mapping DTD 3.0//EN"
"http://hibernate.sourceforge.net/hibernate-mapping-3.0.dtd">
<hibernate-mapping auto-import="false">
    <class name="cdtu.bean.relation.many2many.doubleone2many.Product"
        table="N2N_DOUBLE_1_N_PRODUCT">
        <id name="id" type="java.lang.Integer">
            <column name="ID" />
            <generator class="native" />
        </id>
        <property name="name" type="java.lang.String">
            <column name="NAME" />
        </property>
        <property name="serialNumber" type="java.lang.String">
            <column name="SERIAL_NUMBER" />
        </property>
        <property name="price" type="double">
            <column name="PRICE" />
        </property>
        <property name="stock" type="int">
            <column name="STOCK" />
        </property>
        <set name="orderItems" table="DOUBLE_1_N_ORDERITEM" inverse="true"
            lazy="true">
            <key>
                <column name="PRODUCT_ID" />
            </key>
            <one-to-many class="cdtu.bean.relation.many2many.doubleone2many.OrderItem" />
        </set>
    </class>
</hibernate-mapping>
```

（7）在 cdtu.bean.relation.many2many.doubleone2many 对应包中使用 Hibernate Tools 生成映射文件 Order.hbm.xml，编辑后的代码如下。

```
<?xml version="1.0"?>
<!DOCTYPE hibernate-mapping PUBLIC "-//Hibernate/Hibernate Mapping DTD 3.0//EN"
"http://hibernate.sourceforge.net/hibernate-mapping-3.0.dtd">
<hibernate-mapping auto-import="false">
    <class name="cdtu.bean.relation.many2many.doubleone2many.Order"
        table="N2N_DOUBLE_1_N_ORDER">
```

```
        <id name="id" type="java.lang.Integer">
            <column name="ID" />
            <generator class="native" />
        </id>
        <property name="orderNumber" type="java.lang.String">
            <column name="ORDER_NUMBER" />
        </property>
        <property name="date" type="java.util.Date">
            <column name="DATE" />
        </property>
        <many-to-one name="customer"
            class="cdtu.bean.relation.many2many.doubleone2many.Customer" fetch="join">
            <column name="CUSTOMER_ID" />
        </many-to-one>
        <set name="orderItems" table="DOUBLE_1_N_ORDERITEM" inverse="true"
            lazy="true">
            <key>
                <column name="ORDER_ID" />
            </key>
            <one-to-many class="cdtu.bean.relation.many2many.doubleone2many.OrderItem" />
        </set>
    </class>
</hibernate-mapping>
```

（8）在 cdtu.bean.relation.many2many.doubleone2many 对应包中使用 Hibernate Tools 生成映射文件 OrderItem.hbm.xml，编辑后的代码如下。

```
<?xml version="1.0"?>
<!DOCTYPE hibernate-mapping PUBLIC "-//Hibernate/Hibernate Mapping DTD 3.0//EN"
"http://hibernate.sourceforge.net/hibernate-mapping-3.0.dtd">
<hibernate-mapping auto-import="false">
    <class name="cdtu.bean.relation.many2many.doubleone2many.OrderItem"
        table="N2N_DOUBLE_1_N_ORDERITEM">
        <id name="id" type="java.lang.Long">
            <column name="ID" />
            <generator class="native" />
        </id>
        <many-to-one name="order"
            class="cdtu.bean.relation.many2many.doubleone2many.Order" fetch="join">
            <column name="ORDER_ID" />
        </many-to-one>
        <many-to-one name="product"
            class="cdtu.bean.relation.many2many.doubleone2many.Product" fetch="join">
            <column name="PRODUCT_ID" />
        </many-to-one>
        <property name="quantity" type="int">
            <column name="QUANTITY" />
        </property>
```

```
            <property name="transactionPrice" type="double">
                <column name="TRANSACTION_PRICE" />
            </property>
        </class>
</hibernate-mapping>
```

（9）在 src 根目录下的 Hibernate 配置文件 hibernate.cfg.xml 中，添加下面的映射资源。

```
<!--映射 doubleone2many 资源 -->
        <mapping resource="cdtu/bean/relation/many2many/doubleone2many/OrderItem.hbm.xml" />
        <mapping resource="cdtu/bean/relation/many2many/doubleone2many/Customer.hbm.xml" />
        <mapping resource="cdtu/bean/relation/many2many/doubleone2many/Order.hbm.xml" />
        <mapping resource="cdtu/bean/relation/many2many/doubleone2many/Product.hbm.xml" />
```

（10）在 cdtu.test 子包中新建 Junit 测试类 Doubleone2manyTest.java，添加下面的测试方法。

```
@Test
    public void testAdd() {
        Session session = HibernateSessionFactory.getSession();
        Transaction tx = null;
        Customer customer1 = new Customer("张三", "四川成都", "13900000000");
        Customer customer2 = new Customer("李四", "四川成都", "13900000000");
        Order order1 = new Order("00001", new Date());
        Order order2 = new Order("00001", new Date());
        Product product1 = new Product("Java Web 技术", "100001", 60, 100);
        Product product2 = new Product("Java EE 技术", "100001", 80, 100);
        OrderItem orderItem1 = new OrderItem();
        OrderItem orderItem2 = new OrderItem();
        OrderItem orderItem3 = new OrderItem();
        // 设置关联关系
        customer1.getOrders().add(order1);
        customer1.getOrders().add(order2);
        order1.setCustomer(customer1);
        order2.setCustomer(customer1);

        orderItem1.setOrder(order1);
        orderItem2.setOrder(order1);
        orderItem3.setOrder(order2);
        order1.getOrderItems().add(orderItem1);
        order1.getOrderItems().add(orderItem2);

        orderItem1.setProduct(product1);
        orderItem2.setProduct(product2);
        orderItem3.setProduct(product1);
        product1.getOrderItems().add(orderItem1);
        product1.getOrderItems().add(orderItem2);
        product2.getOrderItems().add(orderItem3);

        try {
```

```
                tx = session.beginTransaction();
                session.save(customer1);
                session.save(customer2);
                session.save(product1);
                session.save(product2);
                session.save(order1);
                session.save(order2);
                session.save(orderItem1);
                session.save(orderItem2);
                session.save(orderItem3);
                tx.commit();
            } catch (Exception e) {
                if (tx != null)
                    tx.rollback();
                throw e;
            } finally {
                HibernateSessionFactory.closeSession();
            }
        }
```

15.6 基于注解的 Hibernate 映射

从 JDK1.5 开始，Java 增加了 Annotation 的支持，通过 Annotation 可以将一些额外的信息写在 Java 源程序中，这些信息可以在编译、类加载、运行时被读取，并执行相应的处理。通过使用 Annotation，程序开发人员可以在不改变原有逻辑的情况下在源文件中嵌入一些补充的信息。

早期 Hibernate 使用 XML 映射文件管理持久化类和数据表之间的映射关系，而 JPA 规范则推荐使用更简单、易用的 Annotation 来管理实体类与数据表之间的映射关系。这样就避免了一个实体需要同时维护两份文件（Java 类和 XML 映射文件），实体类的 Java 代码以及映射信息（写在 Annotation 中）都可集中在一份文件中。

Hibernate 的注解分为类和属性级别注解，下面分别讲解这两种级别注解。

15.6.1 类级别注解

类级别主要包含@Entity 和@Table 两个注解，@Entity 用于映射实体类，@Table 用于映射数据库表。

@Entity(name="tableName")，将一个类声明为一个实体 Bean。name 属性可选，对应数据库中的一个表。若表名与实体类名相同，则可以省略。

@Table(name="",catalog="",schema="")， 此注解可选，通常和@Entity 配合使用，只能标注在实体的 class 定义处，表示实体对应的数据库表的信息。表 15-12 列出了@Table 注解的常用属性。

表 15-12　@Table 注解的常用属性

属性名	功能说明
name	可选，表示表的名称，默认表名和实体名称一致，只有在不一致的情况下才需要指定表名
catalog	可选，表示 Catalog 名称，默认为 Catalog("")
schema	可选，表示 Schema 名称，默认为 Schema("")

15.6.2　属性级别注解

属性注解主要包含：@Id 用于映射生成主键、@Column 用于映射表的列、@Transient 用于定义暂态属性。属性注解细分又可以分为主键相关注解和非主键相关注解，下面将分别讲解这两种注解。

1. 主键相关注解

@Id：定义了映射到数据库表的主键的属性，一个实体只能有一个属性被映射为主键，放置于 getXxxx() 前。

@SequenceGenerator：声明了一个数据库主键生成序列。表@SequenceGenerator 的主要属性如表 15-13 所示。

表 15-13　@SequenceGenerator 的主要属性

属性名	功能说明
name	表示该表主键生成策略名称，它被引用在@GeneratedValue 中设置的“gernerator”值中
strategy	属性指定具体生成器的类名
parameters	得到 strategy 指定的具体生成器所用到的参数

@GeneratedValue：可选注解，用于定义主键生成策略。属性 Strategy 表示主键生成策略，取值如表 15-14 所示。

表 15-14　主键生成策略

属性名	功能说明
GenerationType.AUTO	根据底层数据库自动选择（默认），若数据库支持自动增长类型，则为自动增长
GenerationType.INDENTITY	根据数据库的 Identity 字段生成，支持 DB2、MySQL、MS、SQL Server、SyBase 等数据库的 Identity 类型主键
GenerationType.SEQUENCE	使用 Sequence 来决定主键的取值，适合 Oracle、DB2 等支持 Sequence 的数据库，一般结合@SequenceGenerator 使用（Oracle 使用用 Sequence 实现自动增长）
GenerationType.TABLE	使用指定表来决定主键取值，结合@TableGenerator 使用

2. 非主键相关注解

非主键相关注解主要有@Version、@Basic、@Temporal、@Transient 和@Column 等，下面将分别讲解各自的作用。

（1）@Version：可以在实体 bean 中使用@Version 注解，通过这种方式可添加对乐观锁支持。

（2）@Basic：用于声明属性的存取策略，@Basic(fetch=FetchType.EAGER)表示即时获取（默

认的存取策略)，@Basic(fetch=FetchType.LAZY)表示延迟获取。

（3）@Temporal：用于定义映射到数据库的时间精度。

（4）@Transient：可选，表示该属性并非一个到数据库表的字段的映射，ORM 框架将忽略该属性，如果一个属性并非数据库表的字段映射，就务必将其标示为@Transient，否则 ORM 框架默认其注解为@Basic。

（5）@Column：可将属性映射到列，使用该注解来覆盖默认值，@Column 描述了数据库表中该字段的详细定义。表 15-15 显示@Column 的常用属性。

表 15-15 @Column 的常用属性

属 性 名	功 能 说 明
name	可选，表示数据库表中该字段的名称，默认情形属性名称一致
nullable	可选，表示该字段是否允许为 null，默认为 true
unique	可选，表示该字段是否是唯一标识，默认为 false
length	可选，表示该字段的大小，仅对 String 类型的字段有效，默认值为 255
insertable	可选，表示在 ORM 框架执行插入操作时，该字段是否应出现 INSETRT 语句中，默认为 true
updateable	可选，表示在 ORM 框架执行更新操作时，该字段是否应该出现在 UPDATE 语句中，默认为 true。对于一经创建就不可以更改的字段，该属性非常有用，如 birthday 字段
columnDefinition	可选，表示该字段在数据库中的实际类型。通常 ORM 框架可以根据属性类型自动判断数据库中字段的类型，但是对于 Date 类型仍无法确定数据库中字段类型究竟是 DATE、TIME 还是 TIMESTAMP。此外，String 的默认映射类型为 VARCHAR，如果要将 String 类型映射到特定数据库的 BLOB 或 TEXT 字段类型，则该属性非常有用

15.6.3 注解使用案例

有了基本的注解知识，下面将通过具体的案例讲解注解的基本使用，如果需要更进一步的学习，请参考相关的学习资料。

（1）打开 Eclipse 开发环境，选择 Package Explorer，选中工程 hibernate1 并右击，在弹出的快捷菜单中选择“copy”，然后在 Package Explorer 空白处右击，在弹出的快捷菜单中选择“paste”，然后在弹出的对话框的 Project name 输入框里输入工程名：hibernate5。

（2）编辑 src 的 cap.bean 下的 Product.java 类，编辑后的代码如下。

```
package cdtu.bean;
import javax.persistence.Column;
import javax.persistence.Entity;
import javax.persistence.GeneratedValue;
import javax.persistence.Id;
import javax.persistence.Table;
import org.hibernate.annotations.GenericGenerator;
@Entity
@Table(name = "product_annotation", catalog = "cdtu")
public class Product {
	private Integer id;// 产品 id
	private String name;// 产品名称
```

```
private String serialNumber;// 产品序列号
private double price;// 产品价格
private int stock;// 产品库存量
public Product() {}
public Product(String name, String serialNumber, double price, int stock) {
        this.name = name;
        this.serialNumber = serialNumber;
        this.price = price;
        this.stock = stock;
}
@GenericGenerator(name = "generator", strategy = "increment")
@Id
@GeneratedValue(generator = "generator")
@Column(name = "id", unique = true, nullable = false)
public Integer getId() {
        return id;
}
public void setId(Integer id) {
        this.id = id;
}
@Column(name = "Name", length = 100)
public String getName() {
        return name;
}
public void setName(String name) {
        this.name = name;
}
@Column(name = "SerialNumber", length = 100)
public String getSerialNumber() {
        return serialNumber;
}
public void setSerialNumber(String serialNumber) {
        this.serialNumber = serialNumber;
}
@Column(name = "Price")
public double getPrice() {
        return price;
}
public void setPrice(double price) {
        this.price = price;
}
@Column(name = "Stock")
public int getStock() {
        return stock;
}
public void setStock(int stock) {
        this.stock = stock;
}
```

```
}
```

（3）删除 src 中 cap.bean 下的 Product.hbm.xml，修改 src 下的映射资源为如下的方式。

```
<!--映射 Product 资源 -->
<mapping class="cdtu.bean.Product"/>
```

（4）运行 cap.test 子包 HibernateTest 测试类中的 testAdd 方法，结果和工程 hibernate1 运行相似，唯一的区别是生成的表名不同。

第16章 Hibernate 查询

在前面的章节已经讲解了使用导航对象图的检索方式和根据 OID 的检索方法，在本章中将详细讲解 Hibernate 的三种查询方式：HQL 查询、QBC 查询和本地 SQL 查询。

- HQL 查询：使用面向对象的 HQL 查询语言。
- QBC 查询：使用 QBC（Query By Criteria）API 来检索对象。这种 API 封装了基于字符串形式的查询语句，提供了更加面向对象的查询接口。
- 本地 SQL 查询：使用本地数据库的 SQL 查询语句。

16.1 HQL 查询

Hibernate 的查询语言 HQL（Hibernate Query Language）功能强大，它与 SQL 在语法上非常相似。然而 HQL 是完全面向对象的，其操作的对象是类、对象、属性等。在 HQL 中关键字、函数不区分大小写，但是属性和类名区分大小写。

HQL 基本的查询语法如下。

```
[select attribute_name_list]
from class_name [[as] alias]
[join|left join|right join [fetch]...]
[where ...]
[group by ...]
[having ...]
[order by ...]
```

其中 attribute_name_list 指定查询的属性列表；alias 表示使用别名，前面的 as 可以省略；class_name 指定查询的类名，可以是类的全称。

16.1.1　HQL 检索步骤

使用 HQL 查询的步骤一般包含下面三个步骤：

第一步，通过 Session 的 createQuery()方法创建一个 Query 对象，它包括一个 HQL 查询语句。HQL 查询语句中可以包含命名参数。

第二步，动态绑定参数。

第三步，调用 Query 的 list() 方法执行查询语句。该方法返回 java.util.List 类型的查询结果，在 List 集合中存放了符合查询条件的持久化对象。

HQL 查询语句是面向对象的，Hibernate 负责解析 HQL 查询语句，然后根据对象-关系映射文件中的映射信息，把 HQL 查询语句翻译成相应的 SQL 语句。HQL 查询语句中的主体是域模型中的类及类的属性。

SQL 查询语句是与关系数据库绑定在一起的。SQL 查询语句中的主体是数据库表及表的字段。

16.1.2　HQL 查询案例

下面将通过具体的实例讲解 HQL 的查询使用。

（1）在 Eclipse 开发环境中新建 Java Project 工程：hibernate6。复制工程 hibernate4 中的 cdtu.bean.relation.many2many.doubleone2many 子包下的代码，复制 Hibernate 的配置文件 hibernate.cfg.xml。

（2）在 src 的 cdtu.test 中新建 Junit 测试类 HibernateHQL.java，添加下面的 queryProduct 测试方法，编辑后的代码如下。

```
@Test
	public void queryProduct(){
		session=HibernateSessionFactory.getSession();
		Transaction tx=session.beginTransaction();
		String hql="select p from Product p where p.name like '%Java%'";//第 3 行代码
		List<Product> results=session.createQuery(hql).list();//第 4 行代码
		for(Product product : results){
			System.out.println(product.getId());
			System.out.println(product.getName()+":"+product.getSerialNumber());
			System.out.println(product.getPrice()+":"+product.getStock());
		}
		tx.commit();
		session.close();
	}
```

代码解释：本案例采用拼字符串的方式生成 HQL，第 3 行代码定义了查询的 SQL 语句，第 4 行代码 List<Product> results=session.createQuery(hql).list();是一种链式方法调用，相当于下面两行代码：

```
Query query=session.createQuery(hql);
List<Product> results=query.list();
```

首先通过 session 的 createQuery(hql)方法创建 Query 对象 query，然后通过 query 对象的 list 方法查询结果集。

（3）继续在 HibernateHQL 类中添加 queryProductwithPara 方法，编辑后的代码如下。

```
@Test
    public void queryProductWithPara(){
        session=HibernateSessionFactory.getSession();
        Transaction tx=session.beginTransaction();
        String hql="select p from Product    p where p.name like ?";//第 3 行代码
        List<Product> results= (List<Product>) session.createQuery(hql).setParameter(0, "%Java%").list();
        //第 4 行代码
        for(Product product : results){
            System.out.println(product.getId());
            System.out.println(product.getName()+":"+product.getSerialNumber());
            System.out.println(product.getPrice()+":"+product.getStock());
        }
        tx.commit();
        session.close();
    }
```

代码解释：HQL 中采用“？”来传递参数。第 3 行代码定义 SQL 语句，其中采用“？”来占位需要传递的参数，第 4 行代码通过 Query 对象的 setParameter 方法传递参数，参数的索引位置从 0 开始，然后通过 Query 对象的 list 方法返回查询结果集。

（4）继续在 HibernateHQL 类中添加 queryProductwithPara1 测试方法，编辑后的代码如下。

```
@Test
    public void queryProductWithPara1(){
        session=HibernateSessionFactory.getSession();
        Transaction tx=session.beginTransaction();
        String hql="select p from Product    p where p.name like:pname";//第 1 行代码
        List<Product> results= (List<Product>) session.createQuery(hql).setParameter("pname",
        "%Java%").list();//第 2 行代码
        for(Product product : results){
            System.out.println(product.getId());
            System.out.println(product.getName()+":"+product.getSerialNumber());
            System.out.println(product.getPrice()+":"+product.getStock());
        }
        tx.commit();
        session.close();
    }
```

代码解释：HQL 中采用参数名来传递参数。第 1 行代码在 SQL 语句中定义了占位的参数名为 pname,第 2 行代码同样通过 Query 对象的 setParameter 传递参数，这里是通过指定参数名来传递参数的，而不是索引位置，接着通过 Query 对象的 list 方法返回结果集。

（5）继续在 HibernateHQL 类中添加 queryProductwithParas 测试方法，编辑后的代码如下。

```
@Test
    public void queryProductWithParas(){
```

```
            session=HibernateSessionFactory.getSession();
            Transaction tx=session.beginTransaction();
            String hql="select p from Product  p where p.id in(:name)";
            List<Product> results= (List<Product>) session.createQuery(hql).setParameterList("name", new
Object[]{1,2}).list();
            for(Product product : results){
                System.out.println(product.getId());
                System.out.println(product.getName()+":"+product.getSerialNumber());
                System.out.println(product.getPrice()+":"+product.getStock());
            }
            tx.commit();
            session.close();
        }
```

代码解释：HQL 中采用 setParamterList 方法来传递多个参数。其余的代码和前面的代码类似，不再做具体的解释。

继续在 HibernateHQL 类中添加 queryProductWithPage 测试方法，编辑后的代码如下。

```
    @Test
        public void queryProductWithPage(){
            session=HibernateSessionFactory.getSession();
            Transaction tx=session.beginTransaction();
            String hql="from Product";
            List<Product> results=session.createQuery(hql).setFirstResult(0).setMaxResults(5).list();
            for(Product product : results){
                System.out.println(product.getId());
                System.out.println(product.getName()+":"+product.getSerialNumber());
                System.out.println(product.getPrice()+":"+product.getStock());
            }
            tx.commit();
            session.close();
        }
```

代码解释：Hibernate 中的分页查询，使用 setFirstResult 方法指定查询的起始位置；使用 setMaxResults 方法指定表示每次查询记录数，本例中查询从 0～4 的五条记录数。

（6）继续在 HibernateHQL 类中添加 queryOrderAndCustomer 测试方法，编辑后的代码如下。

```
    @Test
        public void queryOrderAndCustomer(){
            session=HibernateSessionFactory.getSession();
            Transaction tx=session.beginTransaction();
            String hql="select o from Order o";
            List<Order> results=session.createQuery(hql).list();
            for(Order order : results){
                System.out.println("订单号："+order.getId());
                System.out.println("订单的客户：" +order.getCustomer().getName());
            }
            tx.commit();
            session.close();
```

```
}
```

代码解释：根据 Order 对象 order 所对应的客户信息，使用对象导航查询。即通过 order 对象可以查询到相关的客户信息。

（7）继续在 HibernateHQL 类中添加 queryOrderWithJoinCustomer 测试方法，编辑后的代码如下：

```
@Test
    public void queryOrderWithJoinCustomer(){
    session=HibernateSessionFactory.getSession();
    Transaction tx=session.beginTransaction();
    String hql="select  o from Order o inner join fetch o.orderItems order by o.id ";
    List<Order> results=session.createQuery(hql).list();
        for(Order order : results){
            System.out.println("订单号："+order.getId());
            System.out.println("订单的客户：" +order.getCustomer().getName());
        }
        tx.commit();
        session.close();
    }
```

代码解释：在本例中使用 inner join 内连接查询。

16.2 Cretiria 查询

org.hibernate.Criteria 对 SQL 进行封装，即使不了解 SQL 的使用与编写，也可以使用 Criteria 接口所提供的 API 来组合各种查询条件，Hibernate 会自动生成相关的 SQL 查询语句。

16.2.1 QBC 检索步骤

QBC（Query By Cretiria）的检索步骤一般包含下面的三个步骤。

第一步，调用 Session 的 createCriteria 方法创建一个 Criteria 对象。

第二步，设定查询条件。Restrictions 类提供了一系列用于设定查询条件的静态方法，这些静态方法都返回 Criterion 实例，每个 Criterion 实例代表一个查询条件。Criteria 的 add()方法用于加入查询条件。

第三步，调用 Criteria 的 list 方法执行查询语句。该方法返回 List 类型的查询结果，在 List 集合中存放了符合查询条件的持久化对象。

表 16-1 列出了 Restrictions 常用的静态方法。

表 16-1 Restrictions 常用的静态方法

短 语	含 义
Restrictions.eq	等于=
Restrictions.allEq	使用 Map,使用 key/value 进行多个等于的判断
Restrictions.gt	大于>

续表

短 语	含 义
Restrictions.ge	大于等于>=
Restrictions.lt	小于<
Restrictions.le	小于等于<=
Restrictions.between	对应 SQL 的 between 子句
Restrictions.like	对应 SQL 的 like 子句
Restrictions.in	对应 SQL 的 in 子句
Restrictions.and	and 关系
Restrictions.or	or 关系
Restrictions.sqlRestriction	SQL 限定查询

16.2.2 Cretiria 查询案例

下面将通过一个具体的实例讲解 Cretiria 的使用，具有一个直观的、可扩展的条件查询 API 是 Hibernate 的特色。

（1）在 src 的 cdtu.test 中新建 Junit 测试类 HibernateQBC.java，添加下面的 queryByQBC 测试方法。

```
@Test
    public void queryByQBC(){
        session=HibernateSessionFactory.getSession();
        Transaction tx=session.beginTransaction();
        List<Product> results = session.createCriteria(Product.class).setMaxResults(5).list(); //第 3 行代码
        for(Product product : results){
            System.out.print(product.getId());
            System.out.print(product.getName()+":"+product.getSerialNumber());
            System.out.println(product.getPrice()+":"+product.getStock());
        }
        tx.commit();
        session.close();
    }
```

代码解释：第 3 行代码调用 session 的 createCriteria 方法创建 Criteria 对象，然后通过 Criteria 对象的 setMaxResults 方法设置查询为 5 条记录，然后通过 list 方法返回查询的结果集。

（2）在测试类 HibernateQBC.java 中添加下面的 queryByQBCRestriction 测试方法，实现使用 org.hibernate.criterion.Restrictions 类来显示结果集的内容。

```
@Test
    public void queryByQBCRestriction(){
        session=HibernateSessionFactory.getSession();
        Transaction tx=session.beginTransaction();
        List<Product> results = session.createCriteria(Product.class).add(Restrictions.like("name","%Java%")).
list();//第 3 行代码
        for(Product product : results){
```

```
                System.out.print(product.getId());
                System.out.print(product.getName()+":"+product.getSerialNumber());
                System.out.println(product.getPrice()+":"+product.getStock());
            }
            tx.commit();
            session.close();
        }
```

代码解释：第 3 行代码通过调用 Criteria 对象的 add 方法添加限制条件，然后通过 list()方法返回查询的结果集

（3）在测试类 HibernateQBC.java 中添加下面的 queryQBCRestrictionSQL 测试方法，编辑后的代码如下。

```
    @Test
        public void queryQBCRestrictionSQL(){
            session=HibernateSessionFactory.getSession();
            Transaction tx=session.beginTransaction();
            List<Product>  results  =  session.createCriteria(Product.class).add(  Restrictions.sqlRestriction
("{alias}.name like '%java%'")).list();
            for(Product product : results){
                System.out.print(product.getId());
                System.out.print(product.getName()+":"+product.getSerialNumber());
                System.out.println(product.getPrice()+":"+product.getStock());
            }
            tx.commit();
            session.close();
        }
```

代码解释：在 org.hibernate.criterion.Restrictions 类中直接使用 SQL，其中{alias}占位符替换为被查询实体的列别名。

（4）在测试类 HibernateQBC.java 中添加下面的 queryQBCRestrictionProperty 测试方法，编辑后的代码如下。

```
    @Test
        public void queryQBCRestrictionProperty(){
            session=HibernateSessionFactory.getSession();
            Transaction tx=session.beginTransaction();
            List<Product>  results  =  session.createCriteria(Product.class).add(Restrictions.sqlRestriction
("{alias}.name like '%Java%'"))
                        .add(Property.forName("id").in(new Object[]{1,2})).list();//第 3 行代码
            for(Product product : results){
                System.out.print(product.getId());
                System.out.print(product.getName()+":"+product.getSerialNumber());
                System.out.println(product.getPrice()+":"+product.getStock());
            }
            tx.commit();
            session.close();
        }
```

代码解释：第 3 行代码在 org.hibernate.criterion.Restrictions 类中直接使用 SQL，其中使用 Property.forName 方法传递要查询的参数。

（5）在测试类 HibernateQBC.java 中添加下面的 queryByQBCRestriction 测试方法，编辑后的代码如下。

```
@Test
	public void queryQBCByOrder(){
		session=HibernateSessionFactory.getSession();
		Transaction tx=session.beginTransaction();
		List<Product> results = session.createCriteria(Product.class)
				.add( Property.forName("name").like("%Java%") )
				.addOrder( Property.forName("id").asc() )
				.list();	//第 3 行代码
		for(Product product : results){
			System.out.print(product.getId());
			System.out.print(product.getName()+":"+product.getSerialNumber());
			System.out.println(product.getPrice()+":"+product.getStock());
		}
		tx.commit();
		session.close();
	}
```

代码解释：第 3 行代码在 org.hibernate.criterion.Restrictions 类中直接使用 SQL，其中使用 addOrder()方法来对结果集排序。

16.3　本地 SQL 查询

对原生 SQL 查询执行的控制是通过 SQLQuery 接口来进行的，具体是通过执行 Session 对象的 createSQLQuery()方法获取这个接口。可以采用以下代码形式，这个查询指定了 SQL 查询字符串和查询返回的实体。

```
List<Product> results= (List<Product>) session.createSQLQuery(sql).addEntity(Product.class).list();
```

下面将通过一个具体的实例讲解本地 SQL 查询。

（1）在 src 的 cdtu.test 中新建 Junit 测试类 HibernateSQL.java，添加下面的 queryBySQL 测试方法。在该方法中用 addEntity()方法将 SQL 表的别名和实体类联系起来，并且确定查询结果集的形态。

```
@Test
	public void queryBySQL(){
		session=HibernateSessionFactory.getSession();
		Transaction tx=session.beginTransaction();
		String sql="select * from n2n_double_1_n_product";
		List<Product> results= (List<Product>) session.createSQLQuery(sql).addEntity(Product.class).list();
		//第 3 行代码
		for(Product product : results){
```

```
                System.out.println(product.getId());
                System.out.println(product.getName()+":"+product.getSerialNumber());
                System.out.println(product.getPrice()+":"+product.getStock());
            }
            tx.commit();
            session.close();
        }
```

代码解释：第 3 行代码调用 session 对象的 createSQLQuery 创建 SQLQuery 对象，然后通过 SQLQuery 对象的 addEntity 方法指定查询的对象，通过 list 方法返回查询的结果集。

（2）在 Product.hbm.xml 中添加下面的命名 SQL 查询配置：

```
<sql-query name="products">
        <return alias="p" class="cdtu.bean.relation.many2many.doubleone2many.Product" />
        select * from n2n_double_1_n_product as p Where p.NAME LIKE :name
</sql-query>
```

代码解释：<sql-query>使用命名 SQL 查询，在映射文档中定义查询的名字为 products，然后就可以像调用一个命名的HQL查询一样直接调用命名SQL查询，这样就不用调用 addEntity()方法。

（3）在测试类 HibernateSQL.java 中添加下面的 queryBySQLQuery 测试方法。

```
@Test
    public void queryBySQLQuery(){
        session=HibernateSessionFactory.getSession();
        Transaction txn = session.beginTransaction();
        List<Product> results  =  session.getNamedQuery("products").setString("name","%Java%").list();
        //第 3 行代码
        for(Product product : results){
            System.out.print(product.getId());
            System.out.print(product.getName()+":"+product.getSerialNumber());
            System.out.println(product.getPrice()+":"+product.getStock());
        }
        txn.commit();
        session.close();
    }
```

代码解释：第 3 行代码调用 session 的 getNamedQuery 方法获得定义好的 products 命名查询，然后通过 setString 方法传递查询的参数，通过 list 方法返回查询的结果集。

第17章 Struts2+Spring+Hibernate 整合

17.1 Spring 整合 ORM

在持久化技术领域 Spring 的开放性和扩展性得到了一致认可，Spring 不但直接提供了 Spring JDBC，提供了对 JPA、Hibernate、MyBatis 等 ORM 领域的代表者的整合支持，用户完全可以根据需要选择适合自己的技术框架。

Spring 在集成 ORM 框架时，提供了方便的模板类对原 ORM 进行简化封装，以一种更具 Spring 的风格使用 ORM 技术。同时，还可以使用 ORM 框架原生 API 编写程序。

17.2 Spring 中使用 Hibernate

Hibernate 在 ORM 领域具有广泛的影响，拥有广大的使用群体，提供了 ORM 最完整最丰富的实现，本教材使用的 Hibernate 版本为 4.2.8，并且使用原生的 Hibernate API 编写程序实现“增、删、改、查”。

17.2.1 配置 SessionFactory

使用 Hibernate 框架的第一个工作是编写 Hibernate 的配置文件，接着使用这些配置文件实例化 SessionFactory，创建好 Hibernate 的基础设施。

Spring 为创建 SessionFactory 提供了一个好用的 FactoryBean 工厂类，org.springframework.orm.hibernate4.LocalSessionFactoryBean，通过配置一些必需的属性，即可获取一个 SessionFactory 的实例。

Spring 对 ORM 技术的一个重要支持就是提供统一的数据源管理机制，即在 Spring 容器中定义数据源，指定映射文件，设置 Hibernate 控制属性等信息，完成集成组装的工作，并且不使用 hibernate.cfg.xml 配置文件。下面是一个典型的 Spring 中集成 Hibernate 的配置文件。

```
<bean id="dataSource"
        class="org.apache.commons.dbcp.BasicDataSource">
        <property name="driverClassName"
            value="com.mysql.jdbc.Driver">
        </property>
        <property name="url"
            value="jdbc:mysql://localhost:3306/cap">
        </property>
        <property name="username" value="root"></property>
        <property name="password" value="admin"></property>
</bean>
<bean id="sessionFactory"
        class="org.springframework.orm.hibernate4.LocalSessionFactoryBean">
        <property name="dataSource">
            <ref bean="dataSource" />
        </property>
        <property name="hibernateProperties">
            <props>
                <prop key="hibernate.dialect">
                    org.hibernate.dialect.MySQLDialect
                </prop>
                <prop key="hibernate.show_sql">true</prop>
                <prop key="hibernate.hbm2ddl.auto">update</prop>
            </props>
        </property>
        <property name="mappingResources">
            <list>
                <value>cap/bean/Admin.hbm.xml</value>
            </list>
        </property>
    </bean>
```

17.2.2 使用原生的 Hibernate API

通过使用 SessionFactory 的 getCurrentSession()方法能够获取和当前线程绑定的 Session。这一特性使得 Hibernate 自身具备了获取和事务线程绑定的 Session 对象的功能。下面是使用原生 Hibernate API 的示例代码。

```
public class AdminDAOImpl implements AdminDAO {
    private SessionFactory sessionFactory;
    public SessionFactory getSessionFactory() {
        return sessionFactory;
    }
    public void setSessionFactory(SessionFactory sessionFactory) {
        this.sessionFactory = sessionFactory;
```

```
        }
        public Session getSession() {
      return sessionFactory.getCurrentSession();
    }
    }
```

17.2.3 事务处理

Spring 的通用事务管理模型对 Hibernate 完全适用，在本教材中使用基于 aop/tx 的事务管理。下面是典型的事务配置代码。

```
<bean id="txManager"
            class="org.springframework.orm.hibernate4.HibernateTransactionManager">
            <property name="sessionFactory" ref="sessionFactory"/>
    </bean>
    <tx:advice id="txAdvice" transaction-manager="txManager">
            <tx:attributes>
                    <tx:method name="find*" read-only="true" propagation="REQUIRED" />
                    <tx:method name="*"    propagation="REQUIRED"/>
            </tx:attributes>
    </tx:advice>
    <aop:config>
            <aop:pointcut id="transactionPointcut" expression="execution(* cap.dao.impl.*.*(..))" />
            <aop:advisor advice-ref="txAdvice" pointcut-ref="transactionPointcut" />
    </aop:config>
```

17.3 SSH 实现增删改查

有了上面的基础知识，下面将具体讲解 Struts2+Spring+Hibernate 实现“增、删、改、查”。

（1）在 Eclipse 中新建 Dynamic Web Project，工程名为 ssh1，工程的结构图如图 17-1 所示，开发中所需要的开发包如图 17-2 所示。

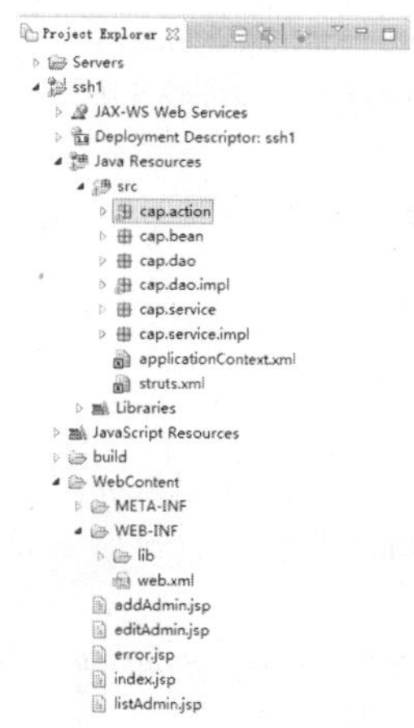

图 17-1 ssh1 的工程结构图

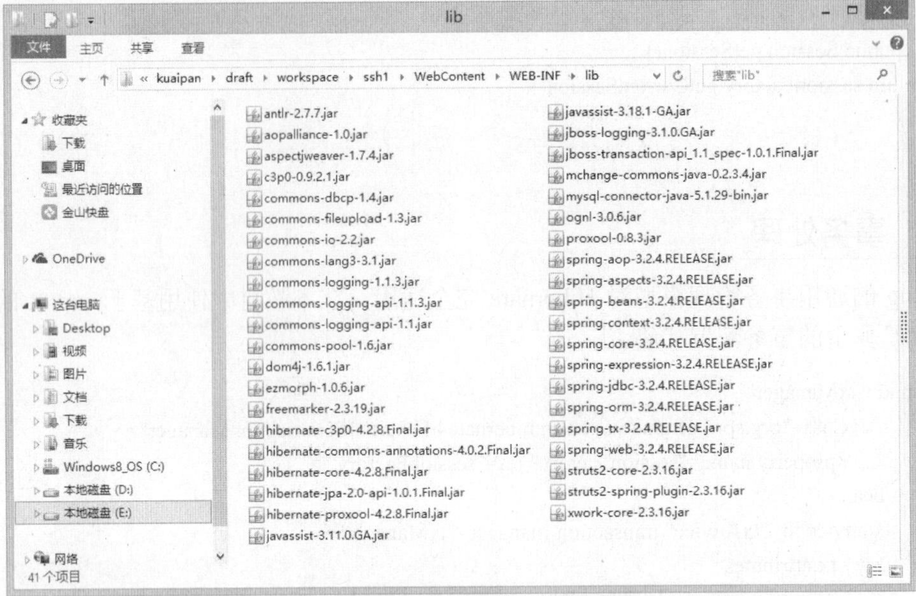

图 17-2　ssh1 工程所需要的开发包

（2）在 cap.bean 中创建持久化类 Admin.java。

```
package cdap.bean;
public class Admin {
        private int id;
        private String username;
        private String password;
      //省略 getters 和 setters

}
```

（3）通过 Hibernate Tools 生成其映射文件 Admin.hbm.xml，代码如下。

```
<?xml version="1.0"?>
<!DOCTYPE hibernate-mapping PUBLIC "-//Hibernate/Hibernate Mapping DTD 3.0//EN"
"http://hibernate.sourceforge.net/hibernate-mapping-3.0.dtd">
<!-- Generated 2015-5-19 21:05:32 by Hibernate Tools 3.4.0.CR1 -->
<hibernate-mapping>
    <class name="cap.bean.Admin" table="ADMIN">
        <id name="id" type="int">
            <column name="ID" />
            <generator class="increment" />
        </id>
        <property name="username" type="java.lang.String">
            <column name="USERNAME" />
        </property>
        <property name="password" type="java.lang.String">
            <column name="PASSWORD" />
        </property>
```

```
        </class>
</hibernate-mapping>
```

（4）在 src 的 cap.dao 中创建 AdminDAO.java 接口，代码如下。

```
package cap.dao;
import java.util.List;
import cap.bean.Admin;
public interface AdminDAO {
        public List<Admin> findAdmins();
        public Admin findById(Integer id);
        public void addAdmin(Admin admin);
        public void deleteAdmin(Integer id);
        public void updtaeAdmin(Admin admin);
}
```

（5）在 src 的 cap.dao.impl 子包中创建 AdminDAO 接口的实现类 AdminDAOImpl.java，编辑后的代码如下。

```
package cap.dao.impl;
import java.util.List;
import org.hibernate.Session;
import org.hibernate.SessionFactory;
import cap.bean.Admin;
import cap.dao.AdminDAO;
public class AdminDAOImpl implements AdminDAO {
private SessionFactory sessionFactory;
        public SessionFactory getSessionFactory() {
                return sessionFactory;
        }
        public void setSessionFactory(SessionFactory sessionFactory) {
                this.sessionFactory = sessionFactory;
        }
        public Session getSession() {
    return sessionFactory.getCurrentSession();
}
        @Override
        public List<Admin> findAdmins() {
                String hql="from Admin m order by m.id";
                return (List<Admin>) getSession().createQuery(hql).list();
        }
        @Override
        public Admin findById(Integer id) {
                return (Admin) getSession().get(Admin.class, id);
        }
        @Override
        public void addAdmin(Admin admin) {
                getSession().save(admin);
```

```
    }
    @Override
    public void deleteAdmin(Integer id) {
        Admin admin=new Admin();
        admin.setId(id);
        getSession().delete(admin);
    }
    @Override
    public void updtaeAdmin(Admin admin) {
        getSession().update(admin);
    }
}
```

（6）剩下的代码 cap.service 子包中的 AdminService.java 接口，以及 cap.service.impl 子包中的 AdminServiceImpl 实现类，cap.action 子包中的 AdminAction.java 类和 10.2 节代码相同，复制代码到各自的包中即可。

（7）将 src 下的 applicationContext.xml 配置文件修改如下。

```
<?xml version="1.0" encoding="UTF-8"?>
<beans xmlns="http://www.springframework.org/schema/beans"
    xmlns:xsi="http://www.w3.org/2001/XMLSchema-instance"
xmlns:aop="http://www.springframework.org/schema/aop"
    xmlns:tx="http://www.springframework.org/schema/tx"
    xsi:schemaLocation="
    http://www.springframework.org/schema/beans
    http://www.springframework.org/schema/beans/spring-beans.xsd
    http://www.springframework.org/schema/tx
    http://www.springframework.org/schema/tx/spring-tx.xsd
    http://www.springframework.org/schema/aop
    http://www.springframework.org/schema/aop/spring-aop.xsd">
    <bean id="dataSource"
        class="org.apache.commons.dbcp.BasicDataSource">
        <property name="driverClassName"
            value="com.mysql.jdbc.Driver">
        </property>
        <property name="url"
            value="jdbc:mysql://localhost:3306/cap">
        </property>
        <property name="username" value="root"></property>
        <property name="password" value="admin"></property>
  </bean>
  <bean id="sessionFactory"
        class="org.springframework.orm.hibernate4.LocalSessionFactoryBean">
        <property name="dataSource">
            <ref bean="dataSource" />
        </property>
        <property name="hibernateProperties">
            <props>
```

```
                <prop key="hibernate.dialect">
                    org.hibernate.dialect.MySQLDialect
                </prop>
                <prop key="hibernate.show_sql">true</prop>
                <prop key="hibernate.hbm2ddl.auto">update</prop>
            </props>
        </property>
        <property name="mappingResources">
            <list>
                <value>cap/bean/Admin.hbm.xml</value>
            </list>
        </property>
    </bean>
<bean id="txManager"
        class="org.springframework.orm.hibernate4.HibernateTransactionManager">
        <property name="sessionFactory" ref="sessionFactory"/>
    </bean>
    <tx:advice id="txAdvice" transaction-manager="txManager">
        <tx:attributes>
            <tx:method name="find*" read-only="true" propagation="REQUIRED" />
            <tx:method name="*"  propagation="REQUIRED"/>
        </tx:attributes>
    </tx:advice>
    <aop:config>
        <aop:pointcut id="transactionPointcut" expression="execution(* cap.dao.impl.*.*(..))" />
        <aop:advisor advice-ref="txAdvice" pointcut-ref="transactionPointcut" />
    </aop:config>
    <bean id="adminDAO" class="cap.dao.impl.AdminDAOImpl">
        <property name="sessionFactory" ref="sessionFactory" />
    </bean>
    <bean id="adminService" class="cap.service.impl.AdminServiceImpl">
        <property name="adminDAO" ref="adminDAO"/>
    </bean>
    <bean id="adminAction" class="cap.action.AdminAction" scope="prototype">
        <property name="adminService" ref="adminService"/>
    </bean>
</beans>
```

（8）WEB-INF 中的配置文件 web.xml 和 src 中的 struts.xml 配置文件与 10.2 节相同，复制相关标签到指定的位置即可。

（9）在 WebContent 目录中新建 listAdmin.jsp 页面，实现代码如下所示。其余的页面 editAdmin.jsp、addAdmin.jsp、error.jsp、以及 index.jsp 和 10.2 节相同，复制到 WebContent 中即可。

```
<%@ page language="java" contentType="text/html; charset=UTF-8"
    pageEncoding="UTF-8"%>
<%@ taglib uri="/struts-tags" prefix="s"%>
<!DOCTYPE html PUBLIC "-//W3C//DTD HTML 4.01 Transitional//EN" "http://www.w3.org/TR/html4/
```

```
loose.dtd">
    <html>
    <head>
    <meta http-equiv="Content-Type" content="text/html; charset=UTF-8">
    <title>显示所有用户</title>
    </head>
    <body>
        <table>
            <s:iterator value="adminList" var="admin">
                <tr>
                    <td><s:property value="#admin.id" /></td>
                    <td><s:property value="#admin.username" /></td>
                    <td><s:property value="#admin.password" /></td>
                    <td><a href="del?id=<s:property value="#admin.id"/>">删除</a></td>
                    <td><a href="edit?id=<s:property value="#admin.id"/>">编辑</a></td>
                </tr>
            </s:iterator>
        </table>
        <a href="addAdmin.jsp">添加用户</a>
    </body>
    </html>
```

（10）在浏览器地址栏中输入：http://localhost:8080/ssh1/list，运行的结果如图 17-3 所示。

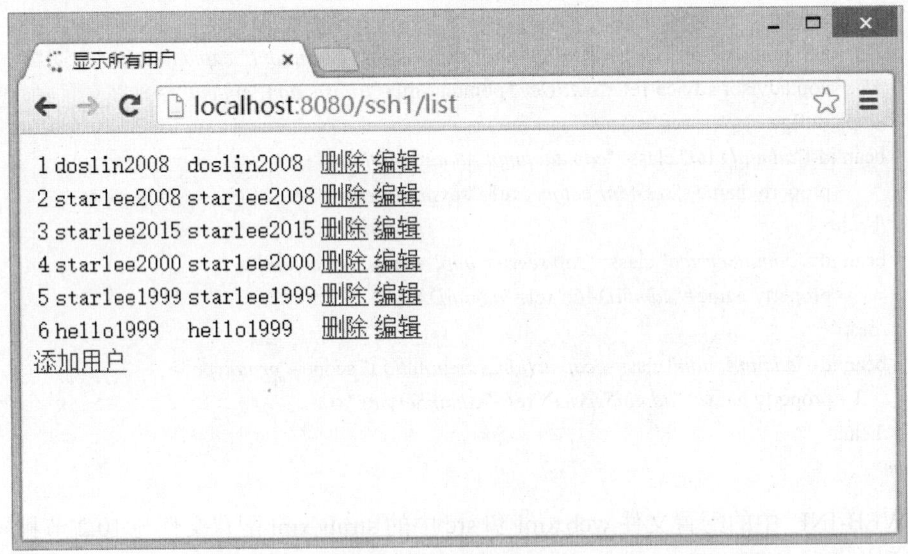

图 17-3　ssh1 的运行结果

17.4　SSH 实现分页

分页显示在 Web 开发中的实现非常重要，下面通过具体的实例讲解 Struts2+Spring+Hibernate 实现分页。

（1）打开 Eclipse 开发环境，选择 Package Explorer，选中工程 ssh1 并右击，在弹出的快捷菜单中选择“copy”；然后在 Package Explorer 的空白处右击，在弹出的菜单中选择“paste”，在 Project name 输入框中输入工程名 ssh2。

（2）在 src 中的 cap.util 新建 PageBean.java 类，编辑后的代码如下。

```
package cdap.util;
import java.util.List;
public class PageBean {
    private List list;
    private int allRow;
    private int totalPage;
    private int currentPage;
    private int pageSize;
    private boolean isFirstPage;
    private boolean isLastPage;
    private boolean hasPreviewPage;
    private boolean hasNextPage;
    public List getList() {
        return list;
    }
    public void setList(List list) {
        this.list = list;
    }
    public int getAllRow() {
        return allRow;
    }
    public void setAllRow(int allRow) {
        this.allRow = allRow;
    }
    public int getTotalPage() {
        return totalPage;
    }
    public void setTotalPage(int totalPage) {
        this.totalPage = totalPage;
    }
    public int getCurrentPage() {
        return currentPage;
    }
    public void setCurrentPage(int currentPage) {
        this.currentPage = currentPage;
    }
    public int getPageSize() {
        return pageSize;
    }
    public void setPageSize(int pageSize) {
        this.pageSize = pageSize;
    }
```

```
    /**
     *初始化分页信息
     */
    public void init(){
        this.isFirstPage = isFirstPage();
        this.isLastPage = isLastPage();
        this.hasPreviewPage = isHasPreviewPage();
        this.hasNextPage = isHasNextPage();
    }
    public boolean isFirstPage() {
        return currentPage == 1;
    }
    public boolean isLastPage() {
        return currentPage == totalPage;
    }
    public boolean isHasPreviewPage() {
        return currentPage !=1;
    }
    public boolean isHasNextPage() {
        return currentPage != totalPage;
    }
    public static int countTotalPage(final int pageSize,final int allRows){
        return allRows%pageSize == 0 ? allRows/pageSize : allRows/pageSize + 1;
    }
    public static int countOffset(final int pageSize, final int currentPage){
        return pageSize*(currentPage-1);
    }
    public static int countCurrentPage(int page){
        return page==0?1:page;
    }
}
```

（3）在 cap.dao 包中的 AdminDAO.java 接口中添加下面的两个方法原型。

```
public List<Admin> findByPage(final String hsql, final int offset, final int length);
public int getTotalCount(String hql);}
```

（4）在 cap.dao.impl 中的 AdminDAO 接口实现类 AdminDAOImpl.java 中添加两个方法的具体实现。

```
@Override
    public List<Admin> findByPage(String hsql, int offset, int length) {
        Query query=getSession().createQuery(hsql);
        query.setFirstResult(offset);
        query.setMaxResults(length);
        List<Admin> list = query.list();
        return list;
    }
```

```
    @Override
    public int getTotalCount(String hql) {
        String count=getSession().createQuery(hql).uniqueResult().toString();
        return Integer.parseInt(count);
    }
```

（5）在 cap.service 包中的 AdminService.java 接口添加下面的方法原型。

```
public PageBean findForPage(int pageSize,int currentPage);
```

（6）在 cap.service.impl 中的 AdminService 接口的实现类 AdminServiceImpl.java 中添加具体的实现方法。

```
@Override
    public PageBean findForPage(int pageSize, int currentPage) {
        final String hsql = "from Admin";
        int allRow = adminDAO.getTotalCount("select count(*) from Admin");
        int totalPage = PageBean.countTotalPage(pageSize, allRow);
        final int offset = PageBean.countOffset(pageSize, currentPage);
        int length = pageSize;
        final int page = PageBean.countCurrentPage(currentPage);
        List list = adminDAO.findByPage(hsql, offset, length);
        PageBean pageBean = new PageBean();
        pageBean.setAllRow(allRow);
        pageBean.setCurrentPage(page);
        pageBean.setPageSize(pageSize);
        pageBean.setTotalPage(totalPage);
        pageBean.setList(list);
        pageBean.init();
        return pageBean;
    }
```

（7）修改 cap.action 包中的 AdminAction.java 类，修改后的代码如下。

```
package cap.action;
import java.util.List;
import cap.bean.Admin;
import cap.service.AdminService;
import cap.util.PageBean;
import com.opensymphony.xwork2.ActionSupport;
public class AdminAction extends ActionSupport {
    private List<Admin> adminList;
    private Integer id;
    private Admin admin;
    private AdminService adminService;
    private PageBean pageBean;
    private int page=1;
    //省略 getters 和 setters

    public String list()
```

```
    {
        pageBean=adminService.findForPage(5, page);
        return SUCCESS;
    }
    public String delete()
    {
        adminService.deleteAdmin(id);
        return SUCCESS;
    }
    public String add()
    {
        adminService.addAdmin(admin);
        return SUCCESS;
    }
    public String edit()
    {
        admin=adminService.findById(id);
        return SUCCESS;
    }
    public String update()
    {
        adminService.updtaeAdmin(admin);
        return SUCCESS;
    }
}
```

（8）修改 WebContent 目录下的 listAdmin.jsp 页面，修改后的代码如下。

```
<%@ page language="java" contentType="text/html; charset=UTF-8"
    pageEncoding="UTF-8"%>
<%@ taglib uri="/struts-tags" prefix="s"%>
<!DOCTYPE html PUBLIC "-//W3C//DTD HTML 4.01 Transitional//EN" "http://www.w3.org/TR/html4/
loose.dtd">
<html>
<hcad>
<meta http-equiv="Content-Type" content="text/html; charset=UTF-8">
<title>显示所有用户</title>
</head>
<body>
    <table>
        <s:iterator value="pageBean.list" var="admin">
            <tr>
                <td><s:property value="#admin.id" /></td>
                <td><s:property value="#admin.username" /></td>
                <td><s:property value="#admin.password" /></td>
                <td><a href="del?id=<s:property value="#admin.id"/>">删除</a></td>
                <td><a href="edit?id=<s:property value="#admin.id"/>">编辑</a></td>
            </tr>
        </s:iterator>
```

```
        <tr>
                <td>总共: <s:property value="pageBean.allRow" /> 条记录
                </td>
                <td>共：<s:property value="pageBean.totalPage" />页
                </td>
                <td>当前第 <s:property value="pageBean.currentPage" />页
                </td>
                <td><s:if test="%{pageBean.currentPage == 1}">
                    第一页 上一页
                </s:if> <s:else>
                        <a href="list.action?page=1">第一页</a>
                        <a
                            href="list.action?page=<s:property value="%{pageBean.currentPage-1}"/>" >
                                                            上一页</a>
                    </s:else></td>
                <td><s:if test="%{pageBean.currentPage != pageBean.totalPage}">
                        <a
                            href="list.action?page=<s:property value="%{pageBean.currentPage+1}"/>"
                                                            >下一页</a>
                        <a
                            href="list.action?page=<s:property value="%{pageBean.totalPage}"/>">最
                                                            后一页</a>
                    </s:if> <s:else>
                    下一页 最后一页
                </s:else></td>
            </tr>
        </table>
        <a href="addAdmin.jsp">添加用户</a>
    </body>
</html>
```

（9）运行工程，在地址栏中输入 http://localhost:8080/ssh2/list，实现的结果如图 17-4 所示。

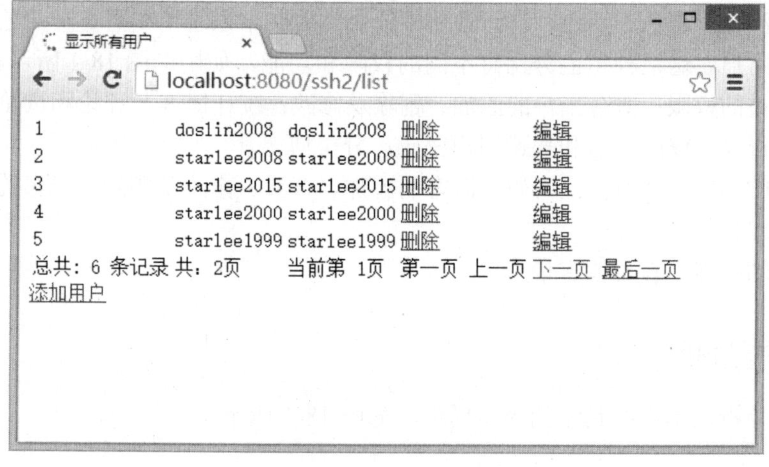

图 17-4　工程 ssh2 的分页显示

第18章 博客系统的设计与实现

18.1 系统需求分析

博客系统，是指使用计算机语言编写，方便用户使用的一个信息交流分享平台。本章节研究的就是在互联网上建立个人博客的一整套系统。

18.1.1 用例图

用例图主要用于描述用例、参与者以及它们之间的关系。用例代表一个完整的功能，UML中的用例是动作步骤的集合。用例与参与者之间是关联关系，这种关联表明哪种参与者或角色能与该用例通信，此关系是双向的一对一关系。

博客管理涉及管理员、博主、访问者三种成员，因此本系统主要有 3 种角色：管理员、普通用户、匿名用户。这三种角色分别有不同的权限和功能，如用例图 18-1 所示。

系统管理员的权限：博客账户的添加、删除及修改；统计博客及评论和浏览次数；博客及评论的删除、查看；修改自己和普通用户密码；分类博客。

普通用户的功能：修改自己密码；博客的添加、查看、修改及删除；博客的搜索；评论的查看及回复管理。

匿名用户的功能：查看博客信息。

18.1.2 功能分析

根据以上分析，本系统具备的功能模块图如图 18-1 所示。

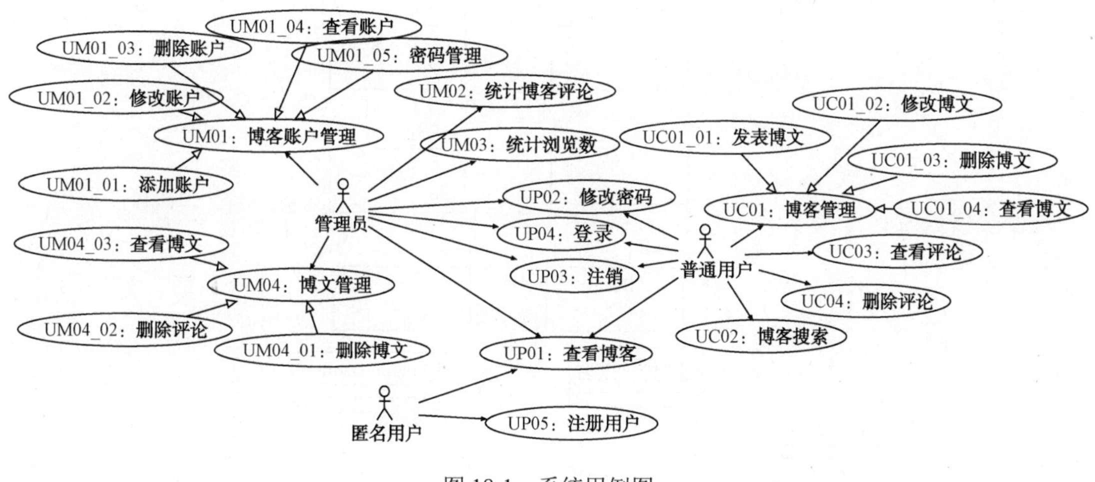

图 18-1　系统用例图

- 系统设置模块

在处理业务之前或系统运行之前一般都要设置一些基础数据，本系统主要包括管理员信息。该信息由系统管理员进行设置，具体管理操作主要包括增、删、改、查、统计、分类等。

- 删除评论

管理员及普通用户可以对博客的评论进行删除。

- 删除博客

管理员及普通用户可以对博客进行删除。

- 查看博客

管理员、普通用户及匿名用户可以对博客的内容、评论、浏览次数及发布日期进行查看。

- 修改密码

系统管理员可以重置所有用户的密码，普通用户能修改自己的密码。

- 博客信息管理

普通用户可以增加、修改博客及增加、回复评论。

18.2　系统架构

本系统的业务逻辑比较简单，容易理解，采用前面章节的 Struts2、Spring、Hibernate 框架整合实现。本系统表示层的技术主要还是采用 JSP 实现，完成数据的显示、接收用户的输入等功能；Hibernate 负责 DAO 层的实现，主要负责对数据库表的操作；Spring 则作为 Bean 的管理容器，主要实现数据源的配置管理，业务 Bean 之间的依赖注入等功能，Struts2 主要负责 MVC 的实现，控制处理流程以及页面的跳转。

通过对功能需求进行分析和概括，绘制出博客信息系统的系统结构图如图 18-2 所示。

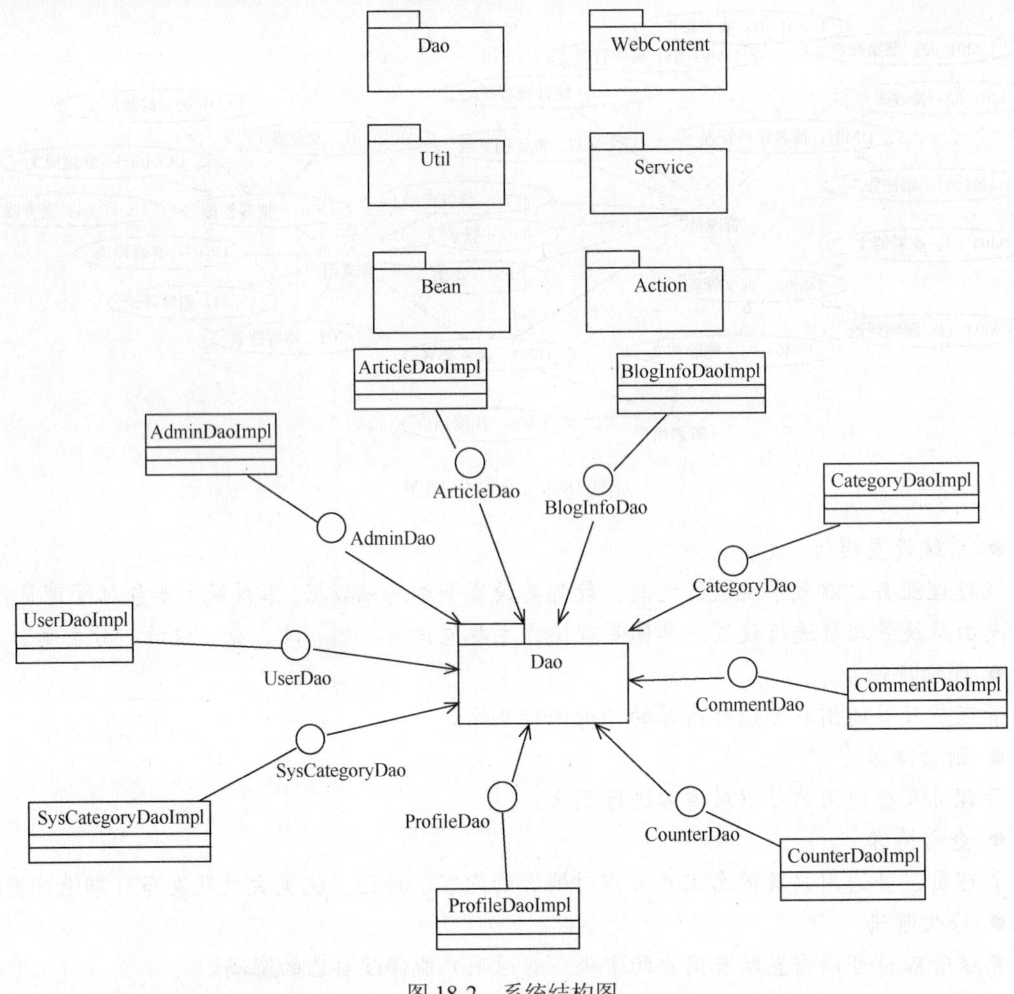

图 18-2　系统结构图

18.3　数据库设计

根据功能描述，本系统有文章信息、评论信息、管理员、普通用户、匿名用户、系统分类信息、用户信息，个人分类信息等实体。一个管理员可以管理多个普通用户，管理员和用户之间是一对多的关系；一名普通用户可以发表多篇文章，普通用户与日志的关系是一对多；一篇文章可以有多个评论，因此文章与评论是一对多的关系；一个普通用户可以发表多个评论，普通用户与评论之间是一对多的关系；一个用户有一条个人信息，用户信息与个人信息是一对一的关系；一个管理员可以建立多个系统分类信息，管理员与系统分类信息是一对多的关系；一个普通用户可以建立多条个人分类信息，普通用户与个人分类信息是一对多的关系；一个系统分类或者一个个人分类可以拥有多篇文章，所以系统分类信息或者个人分类信息与文章信息是一对多的关系。

以上的实体和关系可以通过图 18-3 所示的数据库 E-R 图表示出来。

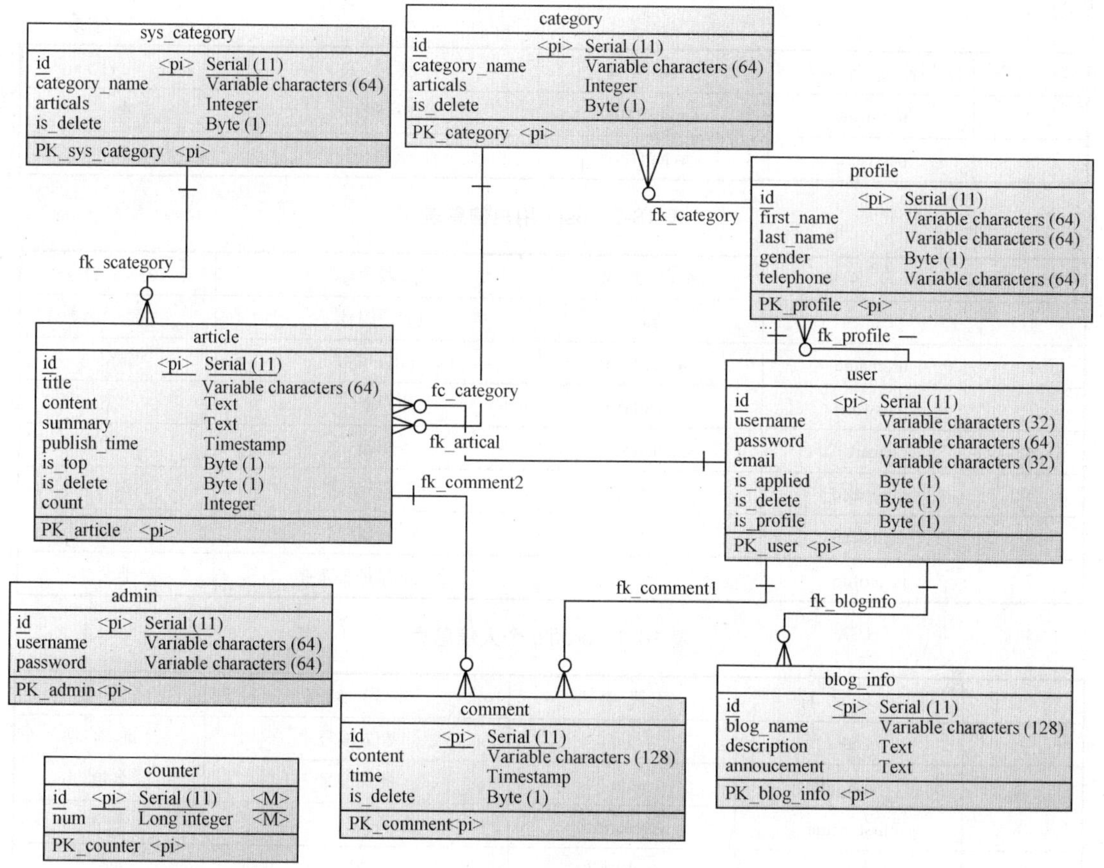

图 18-3 数据库 E-R 图

通过以上的实体关系分析，可设计出数据库的物理模型。本系统使用 MySQL 数据库系统进行存储，需建立 9 张数据表。在设计表字段时，为了编码的方便，字段设计未完全满足第三范式，会有一些字段冗余，如：在设计用户表时，班级编码 classId 为外键，同时也加入了班级名称 className，方便编码。以下为具体设计的 9 张表的情况，见表 18.1～表 18-9。

- 管理员表（admin）：用于存放系统管理员的基本信息。
- 用户信息表（user）：用于存放用户的账号信息。
- 个人信息表（profile）：用于存放用户的详细信息。
- 个人博客信息表（blog_info）：用于存放用户博客的详细信息。
- 系统分类信息表（sys_category）：用于存放系统文章分类信息。
- 个人分类信息表（category）：用于存放用户文章分类信息。
- 博客信息表（article）：用于存放博客文章信息。
- 统计信息表（counter）：用于统计浏览网站的数量。
- 评论信息表（comment）：用于存放文章的评论信息。

表 18-1 admin 管理员表

序　号	字段名称	字段类型	字段含义	备　注
1	id	int	管理员编号	主键、自增

续表

序　号	字段名称	字段类型	字段含义	备　注
2	username	varchar(10)	管理员昵称	非空
3	password	varchar(128)	管理员密码	非空

表 18-2　user 用户信息表

序　号	字段名称	字段类型	字段含义	备　注
1	id	int	用户编号	主键、自增
2	username	varchar(50)	用户名	非空
3	password	varchar(16)	用户密码	非空
4	email	varchar(50)	邮箱	非空
5	is_applied	int	是否已激活	非空
6	is_delete	int	是否已删除	非空
7	is_profile	int	个人信息是否完善	非空

表 18-3　profile 个人信息表

序　号	字段名称	字段类型	字段含义	备　注
1	id	int	个人编号	主键、自增
2	user_id	int	用户编号	外键
3	first_name	varchar(50)	名	非空
4	last_name	varchar(50)	姓	非空
5	gender	int	性别	非空
6	telephone	varchar(50)	电话	非空

表 18-4　sys_category 系统分类信息表

序　号	字段名称	字段类型	字段含义	备　注
1	id	int	系统分类编号	主键、自增
2	category_name	varchar(50)	系统分类名	非空
3	articals	int	分类文章总数	非空
4	is_delete	int	是否删除	非空

表 18-5　category 个人分类信息表

序　号	字段名称	字段类型	字段含义	备　注
1	id	int	个人分类编号	主键、自增
2	user_id	int	用户编号	外键
3	category_name	varchar(50)	个人分类名	非空
4	articals	int	个人文章数量	非空
5	is_delete	int	是否删除	非空

表 18-6　blog_info 博客信息表

序　号	字 段 名 称	字 段 类 型	字 段 含 义	备　注
1	blog_id	int	博客编号	主键、自增
2	User_id	int	用户编号	外键
3	blogName	varchar(100)	博客名	非空
4	description	varchar(1000)	描述	
5	propagate	vachar(500)	宣传	

表 18-7　article 文章信息表

序　号	字 段 名 称	字 段 类 型	字 段 含 义	备　注
1	id	int	文章编号	主键，自增
2	user_id	int	用户编号	外键
3	sys_category_id	int	系统分类编号	外键
4	category_id	int	个人分类编号	外键
5	title	varchar(60)	标题	非空
6	content	mediumtext	内容	非空
7	summary	mediumtext	文章摘要	非空
8	publish_time	timestamp	发表时间	非空
9	is_top	int	是否置顶	非空
10	is_delete	int	是否删除	非空
11	count	int	文章单击数	非空

表 18-8　counter 统计信息表

序　号	字 段 名 称	字 段 类 型	字 段 含 义	备　注
1	id	int	统计编号	主键
2	num	int	统计网站浏览数量	非空

表 18-9　comment 评论信息表

序　号	字 段 名 称	字 段 类 型	字 段 含 义	备　注
1	id	int	评论编号	主键、自增
2	user_id	int	用户编号	外键
3	artical_id	int	文章编号	外键
4	content	varchar(1000)	评论内容	非空
5	time	date	评论时间	非空
6	is_delete	int	是否删除	非空

18.4 配置文件

根据系统的总体设计和数据库设计，首先设计编写 Struts2+Spring+Hibernate 三大框架整合的配置文件以及所需的开发包。工程的页面前端技术采用目前流行的 BootStrap3 实现，读者可以参看《基于 BootStrap3 的 JSP 项目实例教程》自学这部分内容。

（1）在 Eclipse 中创建 Dynamic Web Project 工程，工程名为 blog2，创建好的工程结构图如图 18-4 所示。

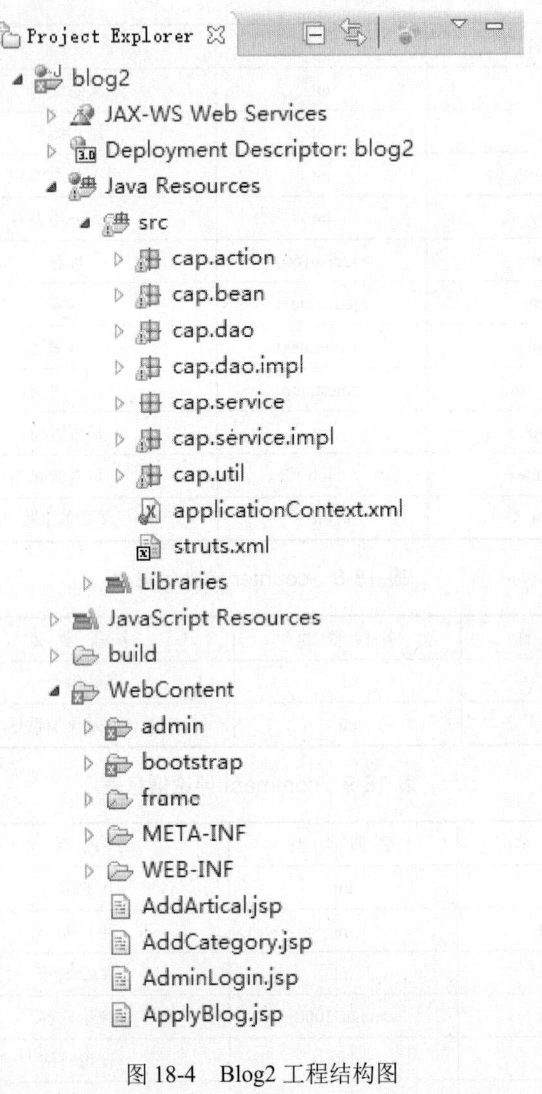

图 18-4　Blog2 工程结构图

（2）修改 WebRoot 目录下 WEB-INF 的 web.xml，编辑后的代码如下，在工程添加 Spring 和 Struts2 框架的支持。

```
<?xml version="1.0" encoding="UTF-8"?>
<web-app   xmlns:xsi="http://www.w3.org/2001/XMLSchema-instance"   xmlns="http://java.sun.com/xml/ns/
```

```
javaee" xsi:schemaLocation="http://java.sun.com/xml/ns/javaee http://java.sun.com/xml/ns/javaee/web-app_3_0.xsd"
id="WebApp_ID" version="3.0">
        <display-name>blog</display-name>
        <welcome-file-list>
          <welcome-file>index</welcome-file>
        </welcome-file-list>
      <context-param>
              <param-name>contextConfigLocation</param-name>
              <param-value>classpath:applicationContext.xml</param-value>
        </context-param>
        <!-- 开启监听 -->
        <listener>
             <listener-class>
                  org.springframework.web.context.ContextLoaderListener
             </listener-class>
        </listener>
        <!-- 配置 OpenSessionInViewFilter,必须在 struts2 监听之前 -->
        <filter>
             <filter-name>lazyLoadingFilter</filter-name>
             <filter-class>
                  org.springframework.orm.hibernate3.support.OpenSessionInViewFilter
             </filter-class>
        </filter>
      <filter>
             <filter-name>struts2</filter-name>
             <filter-class>org.apache.struts2.dispatcher.ng.filter.StrutsPrepareAndExecuteFilter</filter-class>
              <init-param>
                  <param-name>encoding</param-name>
                  <param-value>UTF-8</param-value>
             </init-param>
             <init-param>
                  <param-name>forceEncoding</param-name>
                  <param-value>true</param-value>
             </init-param>
        </filter>
        <filter-mapping>
             <filter-name>struts2</filter-name>
             <url-pattern>/*</url-pattern>
        </filter-mapping>
</web-app>
```

（3）在 src 根目录添加 Spring 的配置文件 applicationContext.xml，编辑后的代码如下。

```
<?xml version="1.0" encoding="UTF-8"?>
<beans
     xmlns="http://www.springframework.org/schema/beans"
     xmlns:xsi="http://www.w3.org/2001/XMLSchema-instance"
     xmlns:p="http://www.springframework.org/schema/p"
     xsi:schemaLocation="http://www.springframework.org/schema/beans  http://www.springframework.org/
```

```
schema/beans/spring-beans-3.0.xsd">
        <bean id="dataSource"
            class="org.apache.commons.dbcp.BasicDataSource">
            <property name="driverClassName"
                value="com.mysql.jdbc.Driver">
            </property>
            <property name="url" value="jdbc:mysql://localhost:3306/blog"></property>
            <property name="username" value="root"></property>
            <property name="password" value="admin"></property>
        </bean>
        <bean id="sessionFactory"
            class="org.springframework.orm.hibernate3.annotation.AnnotationSessionFactoryBean">
            <property name="dataSource">
                <ref bean="dataSource" />
            </property>
            <property name="hibernateProperties">
                <props>
                    <prop key="hibernate.dialect">
                        org.hibernate.dialect.MySQLDialect
                    </prop>
                </props>
            </property>
            <property name="mappingResources">
                <list>
        <!--此处添加 Hibernate 的实体映射文件 hbm.xml -->
    </list>
            </property>
        </bean>
        </beans>
```

（4）在 src 中创建 Struts2 的配置文件 struts.xml，编辑后的代码如下。

```
<?xml version="1.0" encoding="UTF-8" ?>
<!DOCTYPE struts PUBLIC
    "-//Apache Software Foundation//DTD Struts Configuration 2.0//EN"
    "http://struts.apache.org/dtds/struts-2.0.dtd">
<struts>
    <!-- 用户 -->
     <package name="user" namespace="/user" extends="struts-default">
    <!—此处添加 Action 的配置 -->
     </package>
     </struts>
```

（5）添加 Struts2+Spring+Hibernate 整合开发所需要的 jar 包，复制到 WebContent 的 WEB-INF 下的 lib 中，如图 18-5 所示。

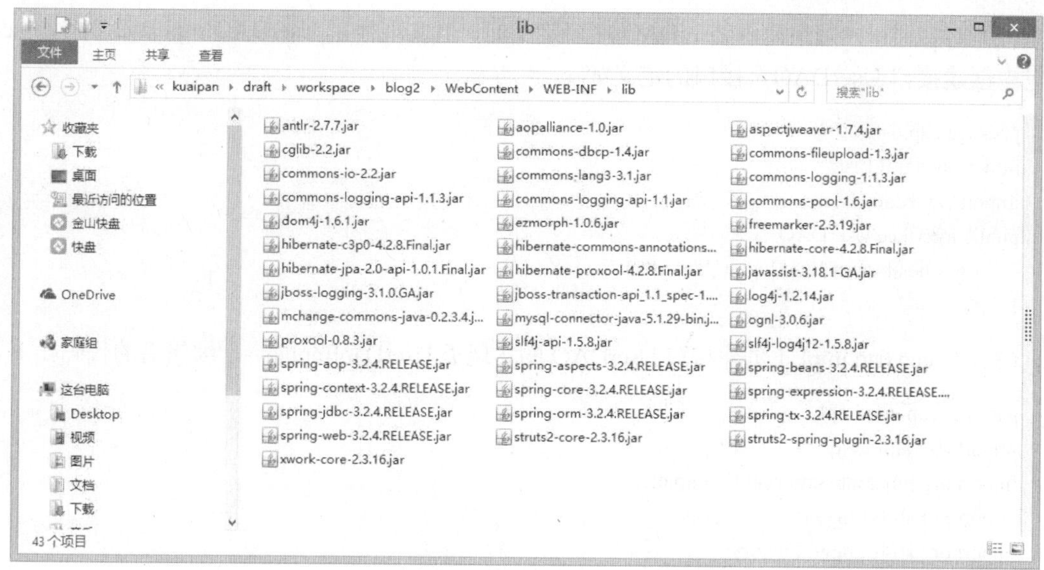

图 18-5　lib 中的 jar 包图

18.5　普通用户模块设计

本系统有 3 类用户：系统管理员、普通注册用户和匿名用户，前两者都需要登录进行相关功能的操作。接下来首先实现用户登录功能。

18.5.1　登录功能

（1）本系统所有用户必须登录之后才能访问页面，登录页面 Login.jsp 的设计如图 18-6 所示。

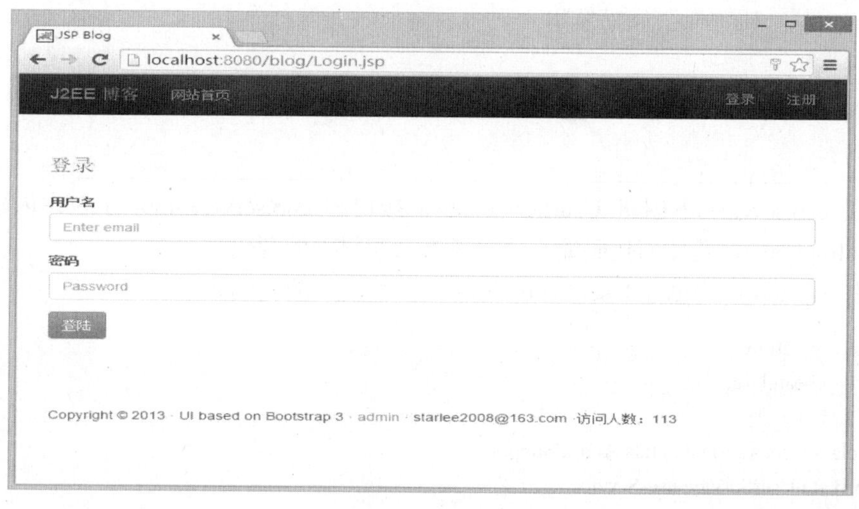

图 18-6　登录页面

（2）对普通用户登录是由 UserDAO 接口定义的，其实现类是 UserDAOImpl.java，在 cap.dao 子包中创建接口 UserDAO，接口的定义如下。

```
package cap.dao;
import java.util.List;
import cap.bean.User;
public interface UserDAO {
	public abstract User login(User user);
}
```

（3）在 cap.dao.impl 下创建接口 UserDAO 的实现类 UserDaoImpl.java，编辑后的代码如下。

```
package cap.dao.impl;
import java.util.List;
import org.hibernate.criterion.Example;
import org.slf4j.Logger;
import org.slf4j.LoggerFactory;
import cap.bean.User;
import cap.dao.UserDAO;
import cap.util.BaseDAO;
public class UserDAOImpl extends BaseDAO implements UserDAO {
//省略部分代码
	@Override
	public User login(User user) {
		User u = null;
		try {
			String hql = "select u from User as u where u.username=? and u.password=?";
			List<User> uList = getSession().createQuery(hql)
					.setParameter(0, user.getUsername())
					.setParameter(1, user.getPassword()).list();
			u = (User) uList.get(0);
		} catch (Exception e) {
			u = null;
			log.error("登录失败");
		}
		return u;
	}
}
```

（4）在 cap.service 下创建 UserService.java 接口和 cap.service.impl 子包中创建其实现类 UserServiceImpl.java，具体的代码编写请参考本书素材中的源码。

（5）在 cap.action 中创建 UserAction.java 类，编辑后的代码如下。

```
package cap.action;
import cap.bean.User;
//省略部分导入类
public class UserAction extends ActionSupport{
	private UserService userService;
	private User u;
```

```
        public UserService getUserService() {
            return userService;
        }
        public void setUserService(UserService userService) {
            this.userService = userService;
        }
        public User getU() {
            return u;
        }
        public void setU(User u) {
            this.u = u;
        }
        public String login(){
            User u1=userService.login(u);
            ActionContext context = ActionContext.getContext();
            Map<String, Object> session = context.getSession();
            if (null != u1) {   //验证成功，还要看 is_delete
                if (u1.getIsDelete() == false) {
                    session.put("user", u1);
                    return SUCCESS;
                } else {
                    session.put("userIsDeleMsg", "该用户已被禁用，无法登录");
                    return ERROR;
                }
            } else {
                session.put("msg", "验证失败，请重新输入用户名或密码！");
                return ERROR;
            }
        }
}
```

18.5.2　文章查看及分页模块

当用户输入正确的用户名和密码之后，进入博客的首页，可以查看不同用户发布的博文，查看界面如图 18-7 所示。

图 18-7　博客查看页面

（1）首先需要在 cap.dao 包中创建 ArticleDAO 接口，添加分页显示的方法，实现的关键代码如下。

```
package cap.dao;
import java.util.List;
import cap.bean.Article;
public interface ArticleDAO {
        public abstract List<Article> getArticleByPage(final int curPage, final int size);
}
```

（2）编写 ArticleDAO 的实现类 ArticleDAOImpl.java，并添加如下实现代码。

```
@Override
public List<Article> getArticleByPage(final int curPage, final int size) {
                final String hql="from Article";
return getSession().createQuery(hql)
                .setFirstResult((curPage-1)*size).setMaxResults(size).list();
}
```

（3）由于要实现分页显示，就在 cap.util 中添加分页类 PageControl.java，实现的代码如下。

```
package cap.util;
import java.util.List;
public class PageControl {
        private int curPage = 1;
        private int pageSize;
        private int totalRows;
        private int totalPages;
        private List list;//用于存放分页数据
        public PageControl(String curPageStr, int totalRows) {
                if (null != curPageStr) {
                        this.curPage = Integer.parseInt(curPageStr); //初始化当前页数
                }
                this.totalRows = totalRows;  //初始化总行数
                this.pageSize=5;//设置每页显示的记录数
                //计算总页数
this.totalPages = (this.totalRows / this.pageSize) + ((this.totalRows % this.pageSize) > 0 ? 1 : 0);
        }
        //省略 getters 和 setters
}
```

（4）接着在 cap.service 子包中创建 ArticleService.java 接口，并添加如下的代码实现分页显示。

```
public interface ArticleService {
        public abstract PageControl getData(String curPageStr);
}
```

（5）在 cap.service.impl 子包中创建 ArticleService 的实现类 ArticleServiceImpl，编辑后的代码如下。

```
package cap.service.impl;
public class ArticleServiceImpl implements ArticleService {
	private ArticleDao artDao;
	public ArticleServiceImpl() {
		artDao=new ArticleDaoImpl();
	}
	@Override
	public PageControl getData(String curPageStr){
		int count=artDao.getAllArtical().size();
		PageControl pc = new PageControl(curPageStr, count);
		List<Article> artList= artDao.getArticleByPage(pc.getCurPage(), pc.getPageSize());
		pc.setList(artList);
		return pc;
	}
}
```

（6）在 cap.action 中创建 IndexAction.java，编辑后的代码如下。

```
package cap.action;
//省略部分导入类
import cap.util.PageControl;
import com.opensymphony.xwork2.ActionSupport;
public class IndexAction extends ActionSupport{
	private static final long serialVersionUID = 1L;
	private ArticleService artService;
	private SysCategoryService scService;
	private CounterService counterService;
	private List<SysCategory> scList ;
	private List<User> ulist;
	private List<Article> artList ;
	private List<Article> tenList;
	private PageControl pc;
	private String curPageStr;
	private long count;
	//省略 getters 和 setters
	public String index(){
		scList=scService.getAllSysCategory();//获取系统分类列表
		ulist=artService.getActiveUser(2); //获取 2 个活跃人数
		tenList=artService.topTenArticle();
		pc=artService.getData(curPageStr);
		count=counterService.getCounter().getNum();
		return SUCCESS;
	}
}
```

（7）在 src 中的 applicationContext.xml 添加下面的 Bean 配置。

```
	<bean id="sessionFactory"
		class="org.springframework.orm.hibernate4.LocalSessionFactoryBean">
```

```
        <property name="dataSource">
            <ref bean="dataSource" />
        </property>
        <property name="hibernateProperties">
            <props>
                <prop key="hibernate.dialect">
                    org.hibernate.dialect.MySQLDialect
                </prop>
            </props>
        </property>
        <property name="mappingResources">
            <list>
            <value>cap/bean/SysCategory.hbm.xml</value>
            <value>cap/bean/BlogInfo.hbm.xml</value>
            <value>cap/bean/Profile.hbm.xml</value>
            <value>cap/bean/User.hbm.xml</value>
            <value>cap/bean/Comment.hbm.xml</value>
            <value>cap/bean/Counter.hbm.xml</value>
            <value>cap/bean/Category.hbm.xml</value>
            <value>cap/bean/Admin.hbm.xml</value>
            <value>cap/bean/Article.hbm.xml</value></list>
        </property>
    </bean>
    <bean id="txManager"
    class="org.springframework.orm.hibernate4.HibernateTransactionManager">
        <property name="sessionFactory" ref="sessionFactory"/>
    </bean>
    <tx:advice id="txAdvice" transaction-manager="txManager">
        <tx:attributes>
            <tx:method name="find*" read-only="true" propagation="REQUIRED" />
            <tx:method name="*"    propagation="REQUIRED"/>
        </tx:attributes>
    </tx:advice>
    <aop:config>
        <aop:pointcut id="transactionPointcut" expression="execution(* cap.dao.impl.*.*(..))" />
        <aop:advisor advice-ref="txAdvice" pointcut-ref="transactionPointcut" />
    </aop:config>
    <bean id="sysCategoryDAO" class="cap.dao.impl.SysCategoryDAOImpl">
        <property name="sessionFactory">
            <ref bean="sessionFactory" />
        </property>
    </bean>
    <bean id="counterDAO" class="cap.dao.impl.CounterDAOImpl">
        <property name="sessionFactory">
            <ref bean="sessionFactory" />
        </property>
    </bean>
    <bean id="articleDAO" class="cap.dao.impl.ArticleDAOImpl">
```

```
        <property name="sessionFactory">
            <ref bean="sessionFactory" />
        </property>
    </bean>

    <bean id="articleService" class="cap.service.impl.ArticleServiceImpl">
    <property name="articleDAO" ref="articleDAO"></property>
    <property name="userDAO" ref="userDAO"></property>
    </bean>

    <bean id="counterService" class="cap.service.impl.CounterServiceImpl">
    <property name="counterDAO" ref="counterDAO"></property>
    </bean>

    <bean id="sysCategoryService" class="cap.service.impl.SysCategoryServiceImpl">
    <property name="scDAO" ref="sysCategoryDAO"></property>
    </bean>

    <bean id="indexAction" class="cap.action.IndexAction" scope="prototype">
    <property name="artService" ref="articleService"></property>
    <property name="scService" ref="sysCategoryService"></property>
    <property name="counterService" ref="counterService"></property>
    </bean>
```

由于 Hibernate 的映射文件和 Spring Bean 的配置会占用大量的篇幅，所以在这里不再一一列举，请读者自行查阅本书素材中的源码。

18.5.3　文章管理模块

当用户单击博客管理后，会弹出下拉菜单，管理页面如图 18-8 所示。

图 18-8　博客管理页面

在创建好的 ArticleAction.java 中，添加下面的方法。

```
public String manage(){
        String curPageStr=curPage;
        ActionContext context = ActionContext.getContext();
        Map<String, Object> session = context.getSession();
        User u=(User) session.get("user");
        pc =artService.getByPageUserId(curPageStr, u.getId());
        return SUCCESS;
}
```

余下的实现代码和前一个模块实现原理相似，唯一的区别就是根据用户的主键查询出文章列表并分页显示。具体的实现请参本随书素材中的源代码。

18.5.4 文章发布模块

单击图中的新建文章，新建文章的页面设计如图 18-9 所示。

图 18-9 发表文章页面

（1）在 cap.action 子包的 ArticleAction.java 类添加下面的实现方法。

```
public String save(){
        ActionContext context = ActionContext.getContext();
        Map<String, Object> session = context.getSession();
        User u=(User) session.get("user");
        art.setUser(u);
        int res = artService.insertArtical(art);
        if (res == 0) {     //添加新文章成功
            session.put("succMsg", "更新文章成功！");
        } else {
```

```
                session.put("errorMsg", "更新文章失败");
            }
            return SUCCESS;
        }
```

（2）在 cap.dao 子包的 ArticleDAO 接口中添加如下的方法，具体的代码如下。

```
public interface ArticleDAO {
    public abstract void save(Article transientInstance);
}
```

（3）在 ArticleDAO 接口中实现类 ArticleDAOImpl.java，实现的关键代码如下。

```
@Override
    public void save(Article transientInstance) {
        log.debug("saving Article instance");
        try {
            getSession().save(transientInstance);
            log.debug("save successful");
        } catch (RuntimeException re) {
            log.error("save failed", re);
            throw re;
        }
    }
```

本节中实现了博客系统的登录、博文的查看、博文的管理、博文的发布。由于篇幅的限制，具体的实现代码没有在此体现完全，只涉及了普通用户模块。此外还有匿名用户，管理用户模块没有讲解，请读者查阅本书素材是的源代码，以便更进一步学习。

Eclipse 开发环境的搭建

要成功运行 Java 及相关工程，首先需要安装 JDK（Java Development Kit），并配置运行 Java 的环境变量，然后再安装 Eclipse IDE 集成开发工具，并在 Eclipse 中配置 Web 服务器 Tomcat。首先讲解 JDK 的环境以及 Java 运行环境变量的设置。

1. Java 环境变量的设置

先根据操作系统版本到 http://java.oracle.com 网站上下载最新版的 Java Development Kit，本教材采用的是 64 位 Windows 8 操作系统版本，下载完成之后进行安装，安装的步骤如下。

（1）双击运行下载的 jdk-8u45-windows-x64，出现 JDK 安装向导，如图 A-1 所示。

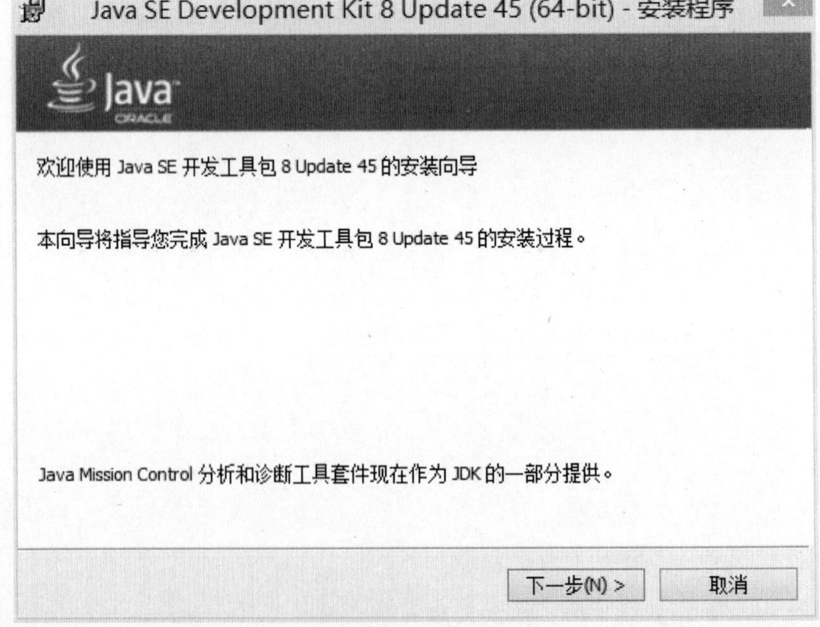

图 A-1　JDK 的安装步骤一

（2）单击“下一步”按钮，选择源代码，将此功能安装在本地磁盘默认驱动器上，如图 A-2 所示。单击“更改”按钮可以设置安装路径。

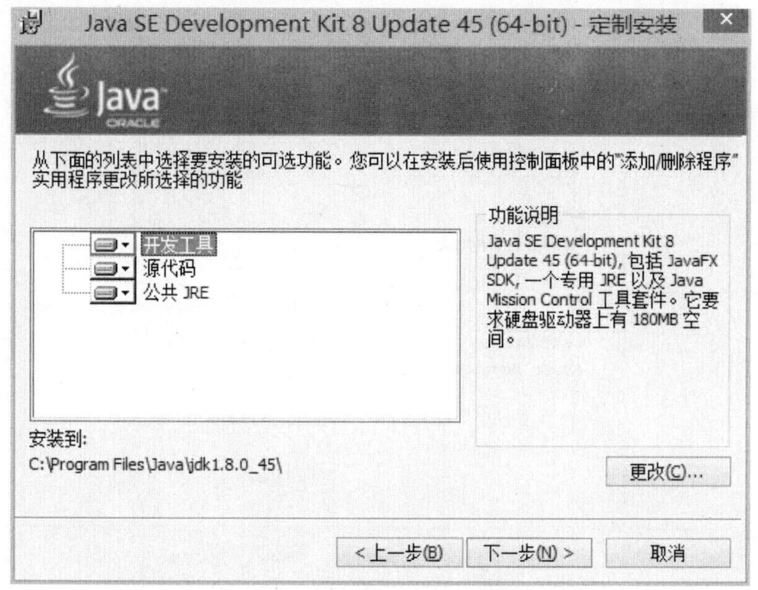

图 A-2　JDK 的安装步骤二

（3）单击“下一步”按钮开始安装 JDK，成功安装将显示如图 A-3 所示界面，单击“关闭”按钮完成 JDK 的安装。

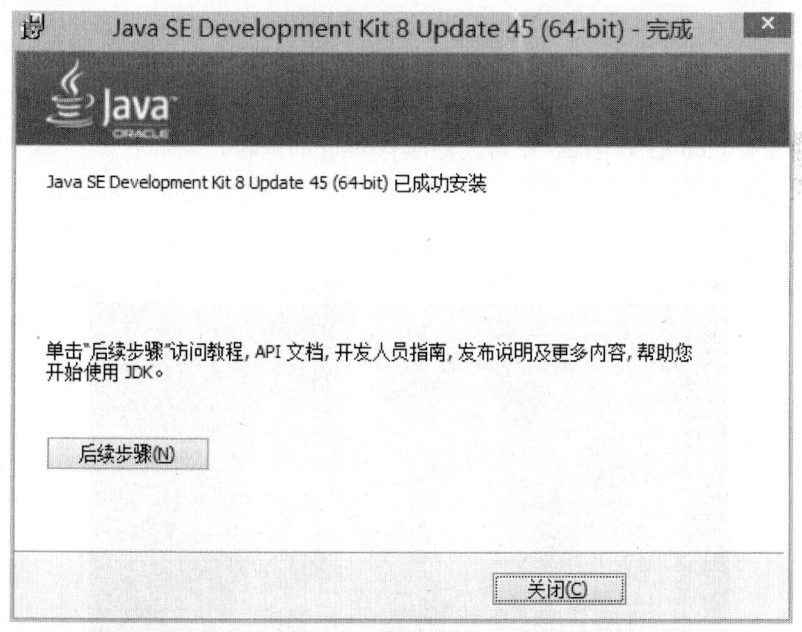

图 A-3　完成 JDK 安装

（4）JDK 安装完成后开始配置 Java 程序设计的环境变量。在 Windows 8 中，打开文件资源管理器，右击“计算机”，选择“属性”，或者通过控制面板→所有控制面板选项→系统→高级系统设置，出现“系统属性”对话框，如图 A-4 所示。

系统属性
计算机名 硬件 高级 系统保护 远程
要进行大多数更改，你必须作为管理员登录。
性能
视觉效果，处理器计划，内存使用，以及虚拟内存
设置(S)...
用户配置文件
与登录帐户相关的桌面设置
设置(E)...
启动和故障恢复
系统启动、系统故障和调试信息
设置(T)...
环境变量(N)...
确定 取消 应用(A)

图 A-4　系统属性

选择“高级”选项卡，单击右下角的“环境变量（N）”按钮，在系统变量增加下面的选项：

① JAVA_HOME: C:\Program Files\Java\jdk1.8.0_45;

② CLASSPATH：.;%JAVA_HOME%\lib\dt.jar;%JAVA_HOME%\lib\tools.jar;

注意：在原有 CLASSPATH 后面增加，不可删除原有 CLASSPATH 的内容。

Path：;%JAVA_HOME%\bin;

注意：在原有 Path 后面增加，不可删除原有 Path 的内容。

（5）验证环境变量是否设置成功：打开命令提示符，输入 javac，如果出现图 A-5 所示内容则表明 Java 环境变量设置成功。

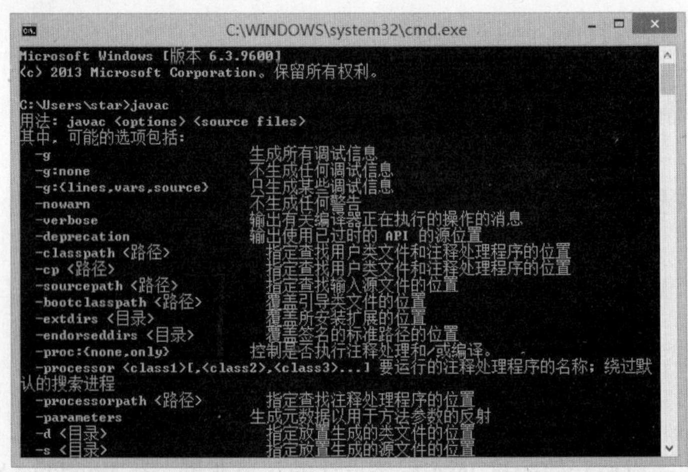

图 A-5　验证环境变量设置是否成功

2. Eclipse 和 Tomcat 集成开发环境配置

本教材所有案例在 Eclipse Mars 版本下编译运行，目前 Tomcat 最新版本为 8.0.24（截止 2015 年 7 月 18 日）。本教材开发案例采用的 Tomcat 的版本号：7.0.63 免安装版。下面将讲解集成开发环境的搭建与配置。

（1）下载 Eclipse。使用浏览器打开 www.eclipse.org ，单击 Download Eclipse，根据操作系统选择 64 位的 Eclipse IDE for Java EE Developers。

（2）下载 Tomcat。在浏览器中打开 http://tomcat.apache.org/，下载 apache-tomcat-7.0.63 免安装版。

（3）解压 Eclipse 和 Tomcat。将下载好的 Eclipse 和 Tomcat 解压到自己的开发环境中：如 D:\JSP，解压 Eclipse 和 Tomcat 到该文件夹下。

（4）进入 Eclipse 目录，双击 Eclipse 启动，启动界面如图 A-6 所示。

图 A-6　Eclipse Mars 启动界面 1

（5）启动完成之后出现如图 A-7 所示界面，关闭该页面之后进入到如图 A-8 所示的界面。

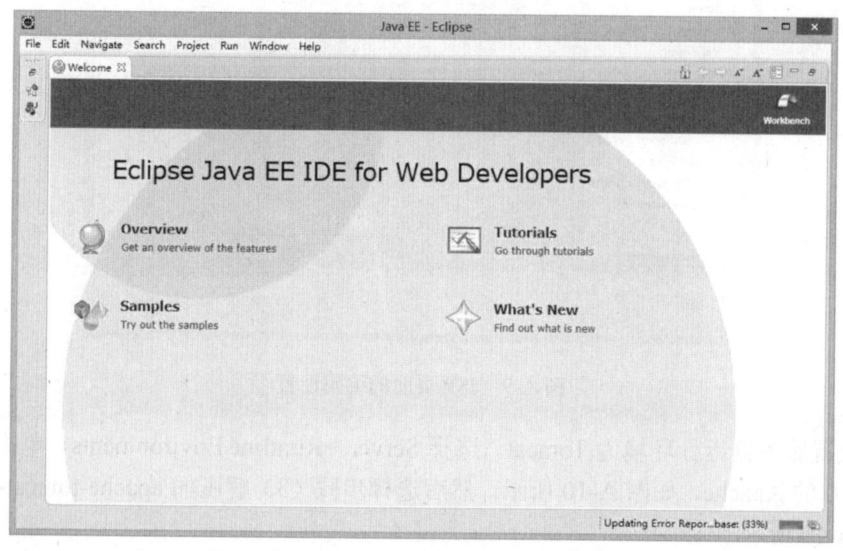

图 A-7　Eclipse 启动界面 2

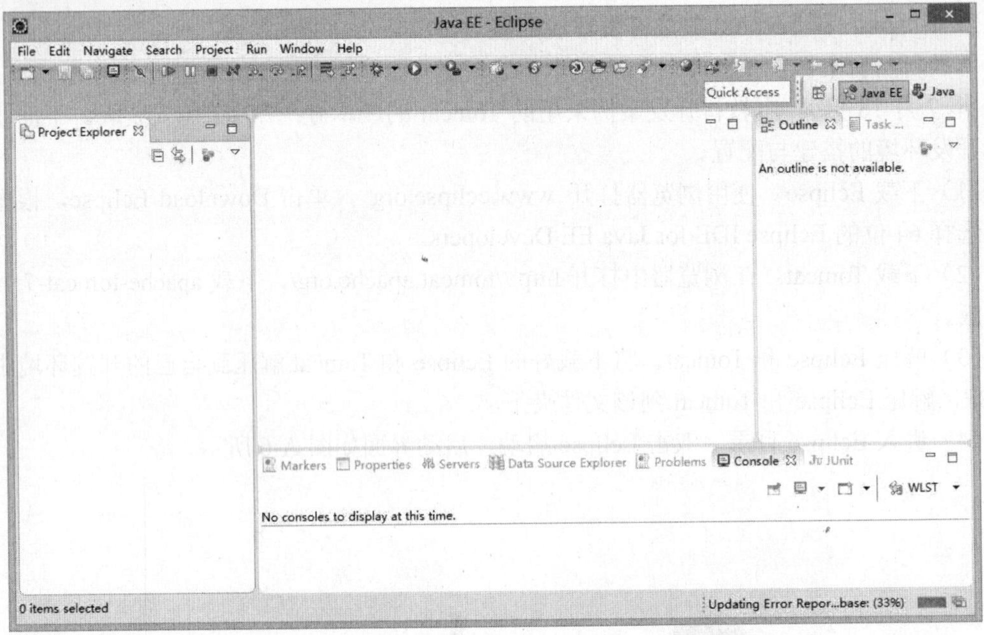

图 A-8　Eclipse 启动界面 3

（6）选择“Window→Prefercences”菜单，出现如图 A-9 所示对话框。选择 Prefercences 下的 Web 进入 JSP Files，将 Encoding 改为 UTF-8。

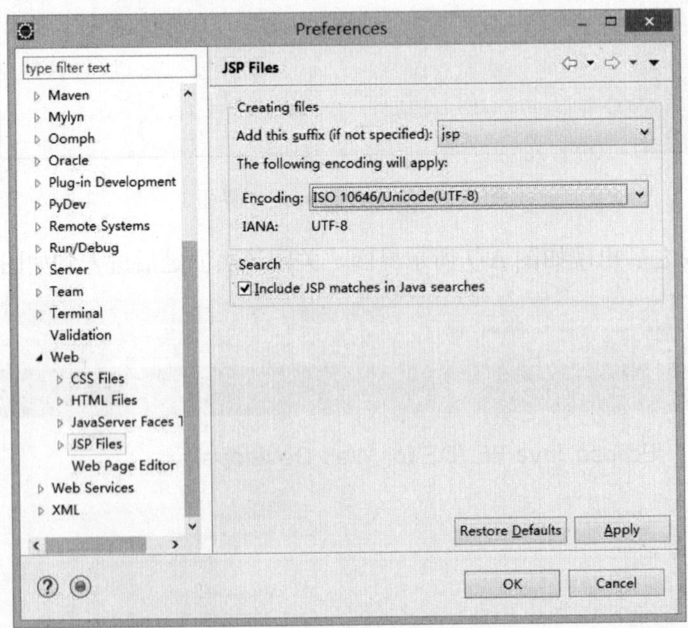

图 A-9　JSP 页面的编码设置

（7）设置服务器运行环境为 Tomcat。选择 Server→Runtime Environments，单击“Add”按钮，添加 7.0 的 Apache，如图 A-10 所示。然后选择步骤（3）解压的 apache-tomcat-7.0.63 的位置，如图 A-11 所示。

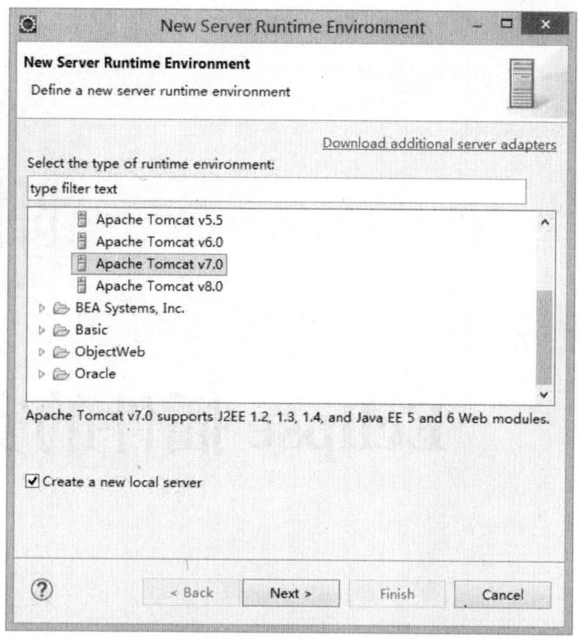

图 A-10　在 Eclipse 中添加 Tomcat 运行环境

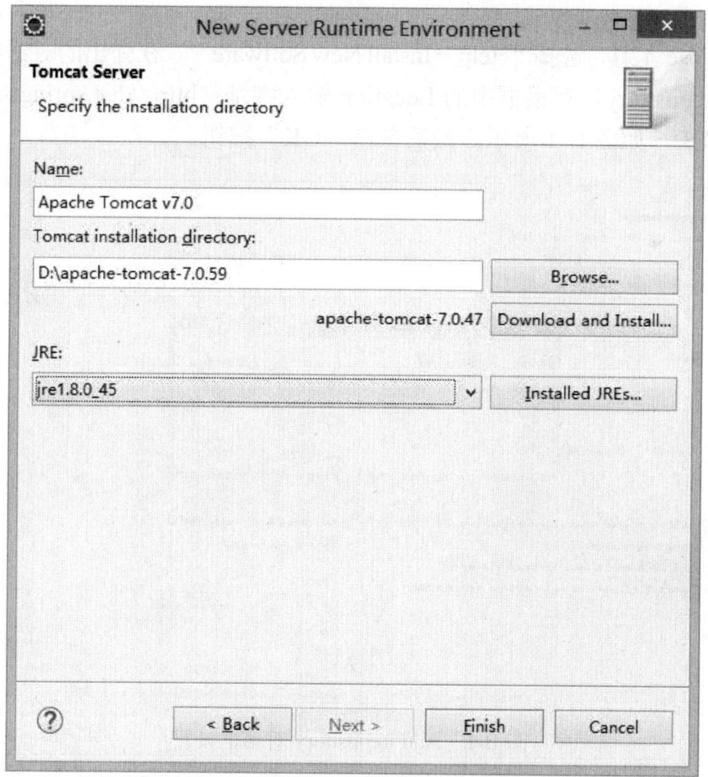

图 A-11　选择 Tomcat 的解压路径

（8）单击“Finish”按钮，Dynamic Web Project 工程开发环境就配置完成。

附录B Eclipse 插件的安装

Eclipse 的插件安装主要有两种方式，一种是在线安装；另一种是下载插件到本地手工安装，下面将以 Spring Tool Suite 插件的安装为例进行讲解。

1. 在线安装

（1）打开 Eclipse 工具，选择“Help→Install New Software”，在弹出的对话框中单击“Add”按钮，在“Add Repository”对话框中的 Location 输入网址：http://dist.springsource.com/release/TOOLS/ update/e4.3/，如图 B-1 所示，最后单击“OK”按钮。

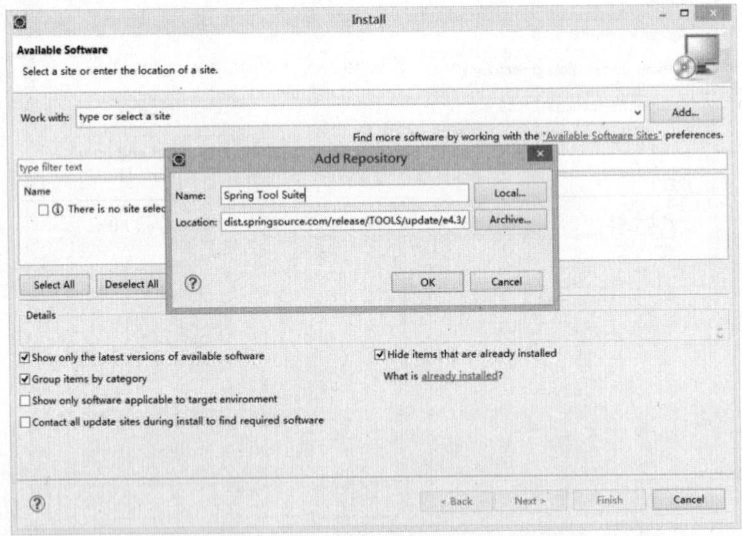

图 B-1　Add Repository 对话框设置

（2）选择需要安装的插件，选择安装带“/Spring IDE”的插件。取消勾选“Contact all update sites during install to find required software”复选框，如图 B-2 所示。

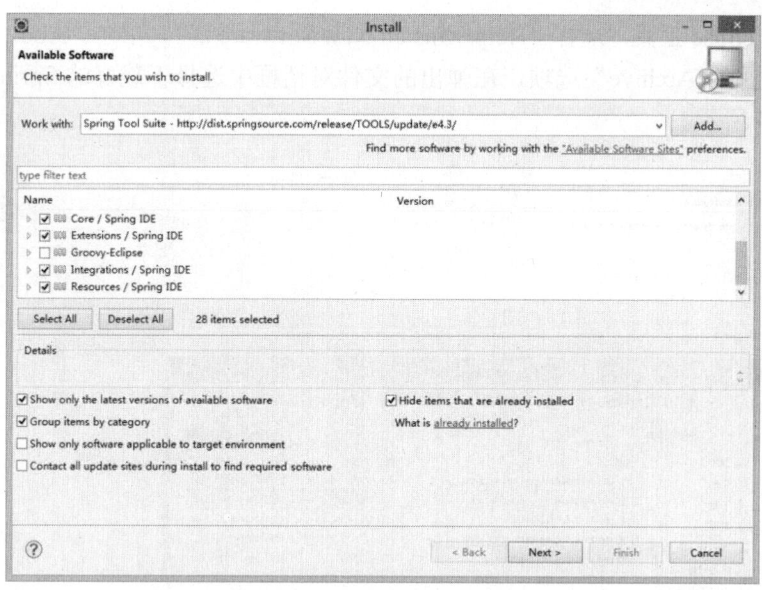

图 B-2　插件选择安装图

（3）单击两次“Next”按钮之后接受安装 Spring Tool Suite 的安装协议，如图 B-3 所示。选择“I accept the terms of the licene agreements”单选按钮。单击“Finish”按钮开始安装，在安装的过程中会出现 Security Warning 的警告框（由于安装的插件包含没有签名的内容），单击“OK”按钮继续安装。按提示完成安装，重启 Eclipse IDE 即可。

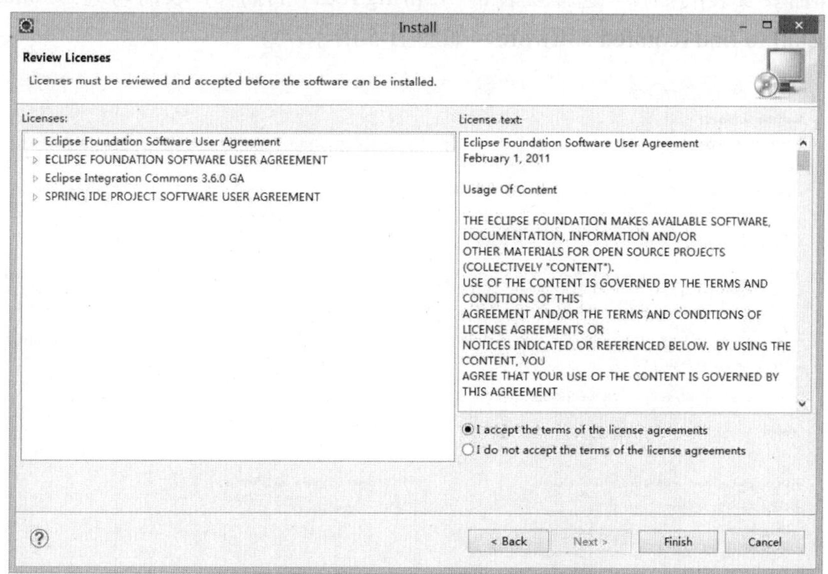

图 B-3　安装 Spring Tool Suite 步骤图

2. 手动安装

（1）通过网络下载需要的插件，比如这里要安装的 Spring Tool Suite 插件的网址为：http://spring.io/tools/sts/all，根据 Eclipse 的版本下载需要的插件离线安装包，然后解压到本地磁盘中。

（2）打开 Eclipse 工具，选择“Help→Install New Software...”，单击“Add…”按钮，在弹出的对话框中选择“Archive”选项，在弹出的文件对话框中选择下载好的 Spring Tool Suite 的路径，如图 B-4 所示。

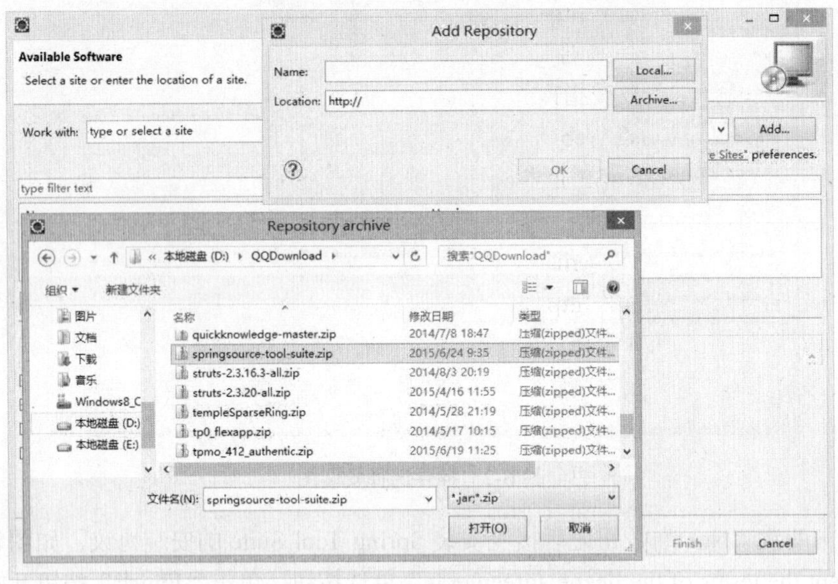

图 B-4　选择 JBoss Toos 步骤图

（3）选择需要安装的插件，选择安装带“/Spring IDE”的插件。取消勾选：Contact all update sites during install to find required softwore，如图 B-5 所示。

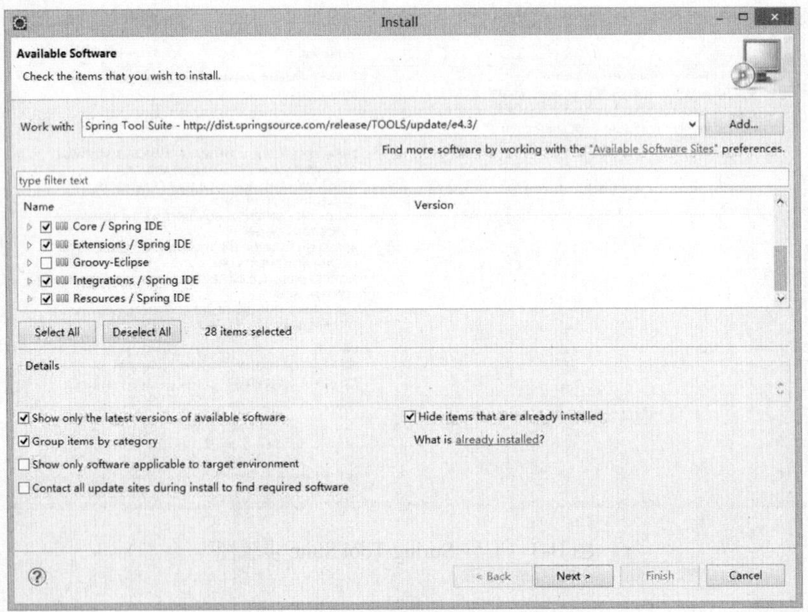

图 B-5　选择需要安装的插件

（4）其余的步骤和在线安装一样，最后重启 Eclipse IDE 即完成插件的安装。

参 考 文 献

[1] Struts2 Documentation，http://struts.apache.org/.

[2] 布朗，戴维斯等著，马召等译．Struts2 in Action．北京：人民邮电出版社．2010 年．

[3] 李刚编．Struts2.x 权威指南．北京：电子工业出版社．2012 年．

[4] 杨少敏编著．Struts2 Web 开发学习实录．北京：清华大学出版社．2011 年．

[5] 李刚编．轻量级 JavaEE 企业应用实战 Struts2+Spring3+Hibernate 整合开发．北京：电子工业出版社．2011 年．

[6] 蒲子明等编. Struts2+Hibernate+Spring 整合开发技术详解. 北京: 清华大学出版社. 2010 年.

[7] Spring Documentation，http://www.spring.io/docs.

[8] Craig Walls 著，耿渊等译．Spring 实战．北京：人民邮电出版社．2013 年．

[9] 陈雄华编著．Spring3.0 就这么简单．北京：人民邮电出版社．2013 年．

[10] 孙卫琴编著．Hibernate 逍遥游记．北京：电子工业出版社．2012 年．

[11] Hibernate Documentation，www.hibernate.org.

[12] 李刚编著．疯狂 Java 讲义．北京：电子工业出版社．2008 年．

[13] 李明欣、林琳等编．基于 BootStrap3 的 JSP 项目实例教程．北京：航空航天大学出版社，2015 年．

反侵权盗版声明